DATE DUE

Forestry for Human Development

Hans Kristian Seip

Forestry for Human Development

A Global Imperative

SCANDINAVIAN UNIVERSITY PRESS
Oslo – Stockholm – Copenhagen – Oxford – Boston

Scandinavian University Press (Universitetsforlaget AS)
P.O. Box 2959 Tøyen, N-0608 Oslo, Norway
Fax +47 22 57 53 53
URL: http:/www.scup.no

Stockholm office
SCUP, Scandinavian University Press
P.O. Box 3255, S-103 65 Stockholm, Sweden

Copenhagen office
Scandinavian University Press AS
P.O. Box 54, DK-1002 København K, Denmark

Oxford office
Scandinavian University Press UK
60 St. Aldates, Oxford OX1 1ST, England

Boston office
Scandinavian University Press North America
875 Massachusetts Ave., Ste. 84, Cambridge MA 02139, USA
Fax +1 617 354 6875

© Scandinavian University Press (Universitetsforlaget AS), Oslo 1996
ISBN 82-00-22761-8

Published with a grant from The Royal Ministry of Agriculture, Norway

Design: Astrid Elisabeth Jørgensen
Cover illustration: © The Image Bank/Steve Krongard
Typeset in 11.5 on 13 point Photina by
Heien Fotosats A.s, Norway
Printed on 100 gms Partner Offset ∞ by SRT A.s, Norway

Foreword

The forest is one of the most important vegetation communities in the world, and the management of the world's forests has always been of great importance for the welfare of mankind. Today, we use forest resources as firewood and energy for heating and cooking, as building materials for our homes, as important elements in larger construction projects, and, ever increasingly, as paper products. The forest and its biomass are essential to the global storage of carbon and thus critical in the struggle against global warming. In addition, the forest is for many people an important resource for recreational activities. And last but not least, the forest is a living society and a sanctuary for a great part of the biological diversity on this globe, which in turn is decisive for ecological balance.

Given the great value of the forests, it is understandable that the management of these areas and resources is highly disputed. Nonetheless, it is clear that we must arrive at principles for the management of forest resources that satisfy not only national but also global concerns.

With his experience as a scientist, a prominent senior official, and a politician, Hans Kristian Seip has the ideal foundation for contributing to this work. He has also for many years been engaged as a high-level advisor in forestry matters all over the world, both within the UN system and for numerous national governments.

To everybody concerned about sustainable development and long-range resource management, this book offers valuable information and views, and we are thankful for being able to contribute financially to the publication.

Oluf Aalde

Director General of Forestry
Royal Norwegian Ministry of Agriculture

Oslo, August 1996

Contents

Acknowledgements

The idea of attempting to give a brief overview of the problems encountered in managing forest resources for a multitude of purposes came to me as I was working on an ITTO pre-project in Sarawak, Malaysia. The project I was given the task of formulating was called "Sustainable Multiple Use of Hill Forest in Sarawak, Malaysia". Studying the problems and possibilities in Sarawak and the valuable discussions I had with relevant people, in particular the Forest Department staff, were very inspiring and enlightening. For this I am very grateful. Among the many people who gave me good advice I would particularly like to mention Cheong Ek Choon and Joseph Jawa Kendawang, who both read parts of the manuscript for this book and made important comments.

I have also accumulated useful experiences from other parts of the world where for a couple of decades I have been doing occasional jobs for various organizations. Contact with the Food and Agriculture Organization of the United Nations during this time meant a great deal to me. Again a lot of people deserve a mention. Let me here just mention Jean-Paul Lanly and Klaus Janz, who gave me access to the Global Synthesis of the 1990 Forest Resources Assessment even before it was publicly available.

Even if the scope of the book is global and with more emphasis on the tropics than on the temperate zone, much of the thinking will be rooted in the Nordic professional tradition, in particular that of my own country, Norway. To make sure that these roots were utilized appropriately I asked three of my friends and colleagues to go through the manuscript and criticize it. These three, Ivar Ekanger, Torstein Opheim, and Berit Sanness, have in various ways been of great help, and I am very grateful to them.

Even my own family has helped me. My son Henrik and his wife, Laila, have introduced me to the mysteries of a PC. My daughter Ellen gave valuable advice on the structure of the book, and my wife, Anne-Marie, has joined me around the world and encouraged me to put the whole thing together.

I thank everybody for inspiration and help.

Hans Kristian Seip
Ås, April 1995

Abbreviations

CIB	Comité International du Bois
CIFOR	Center for International Forestry Research
DBH	diameter at breast height
ECE	United Nations Economic Commission for Europe
ETTS	*European Timber Trends and Prospects*
FAO	Food and Agricultural Organization of the United Nations
FOWL	forest and other wooded land
GDP	gross domestic product
ICRAF	International Council for Research on Agro-Forestry
IIED	International Institute for Environment and Development
INGO	international non-governmental organization
IPCC	Intergovernmental Panel on Climate Change
ITTO	International Tropical Timber Organization
IUCN	International Union for the Conservation of Nature and Natural Resources
IUFRO	International Union of Forest Research Organizations
m³ EQ	cubic metres, equivalent volume of wood in the raw
NAI	net annual increment
NFA	national forest authorities
NGO	non-governmental organization
ODA	Official Development Assistance

PPP	purchasing power parity
TFA	Tropical Forestry Action Programme
UCB	Union pour le Commerce des Bois Tropicaux dans la CEE
UN	United Nations
UNCE	United Nations Conference on Environment and Development
UNCTAD	United Nations Conference on Trade and Development
UNDP	United Nations Development Programme
UNEP	United Nations Environmental Programme
Unesco	United Nations Educational, Scientific, and Cultural Organization
WCED	World Commission on Environment and Development
WWF	World Wide Fund for Nature

Some basic definitions

Forest   An ecosystem where trees grow to more than about 7 metres in height and are able to produce wood. The crown cover (stand density) is usually more than 20 percent of the area (in some countries, 10 percent).

Forestry   The activity concerned with establishing, managing and utilising forest ecosystems, aiming at satisfying human needs.

Deforestation   A change of land use from forest to other land use, or permanent depletion of forest crown cover to less than 10 percent. Temporary clear-cutting as an instrument in forest management is not deforestation.

Degradation   As distinct from deforestation, "this means that the production potential of the forest area is reduced through outside, mostly human, factors. Degradation can include damage both to the soil and to the growing stock" (Bostrøm, 1986).

Sustainable development   "[D]evelopment that meets the needs of the present without compromising the ability of future generations to meet their own needs" (WCED, 1987a).

Introduction

> For, contrary to what many outsiders believe, forestry is not in its essence,
> about trees. It is about people. It is about trees only so far as they can serve
> the needs of people.
>
> (Jack Westoby, 1987)

Forests and forestry have very much come into public focus during
the last half of the twentieth century, for very legitimate reasons.
It is the result of a development in which scarcity or a fear of scarcity of the services expected from the forests has been recognized.

These services cannot easily be measured by the same scale. In
this presentation it is assumed that human welfare is what we are
aiming at. In a global setting this is not limited to any particular
group, but applies to the general welfare of humanity now and in
the future. It corresponds to the term "sustainable development"
used by the World Commission on Environment and Development
(the Brundtland Commission). Forest management that aims at
such a goal is in practical terms not found by straightforward
calculation. Nor can it be definitively established in a textbook or a
declaration, however important.

This book is an attempt to offer a brief description of the current
situation and to establish some lines of thinking that might be
helpful in the ongoing discussion.

To attack the problem it has been felt useful to take a look at the
reasons why we have arrived at this rather confusing situation.
This has been done in Chapter 1, which briefly presents the histor-

ical background. Chapter 2 gives a (similarly brief) overview of global forest resources.

In Chapter 3 a number of important services expected from the forests are discussed, and Chapter 4 takes up the problem of to what degree these services can be produced simultaneously.

Chapter 5 suggests some possibilities for increasing the compatibility between various services, and Chapter 6 tries to point to a path through the thicket of problems at the national level. The recognition of international responsibility is emphasized as a prerequisite for finding solutions near the optimal aggregate benefit of forests.

The manuscript was submitted in April 1995.

Chapter 1

Historical background

1.1. Snapshots from forestry before World War II

The history of mankind's use of forests is as old as the history of mankind. Hunting and gathering were basic for livelihood. Food, fuel, crude weapons, shelter, and even medicine were early benefits harvested from the forest. As agriculture developed for food production, the forest would sometimes be regarded as an enemy that had to be fought against to yield space for agricultural crops. It was also felt to be dark, dangerous, and mystical, so that cutting down and burning the forest were looked upon as positive activities.

Even 400–500 years BC, deforestation in some Mediterranean regions was so widespread that regulations were introduced by law. Reforestation was also tried. The need for wood and for agricultural land was, however, so pressing that regulations were not followed up. Much later, some laws concerning forests were passed in central Europe, but these laws were concerned only with property boundaries and who had the right to use what.

From the year AD 1224 we find an interesting example of "long-term planning" (Grøn, 1945). The forest area of the Monre estate – owned by the bishop – near Langensalza in Mainz (Germany) needed regulation to secure future requirements of wood. It was achieved in a way that was very simple but probably sufficient for the purpose. The area was divided into a number of lots equal to the number of years estimated to be a useful rotation (say, 80 years). A different lot was cut each year, until a new round could be started 80 years in the future. This was clearly a

← Shadow-tree in silvo-pasture, Ethiopia. © Hans Kristian Seip.

plan for sustainable management, and the focus seems to have been on wood that was threatened by scarcity, even if other benefits (wildlife, water) may have been involved.

In subsequent centuries several similar plans were worked out in central Europe. More sophisticated methods were developed to improve accuracy, but always focusing on wood.

Sustainability has been a leading principle in forestry for as long as deliberate management of forest resources has existed. It was clearly expressed by Hartig in Germany in 1804: "Aus den Waldungen des Staates soll jährlich nicht mehr und nicht weniger Holz genommen werden, als bei guter Bewirtschaftung mit immerwärender Nachhaltigkeit daraus zu beziehen möglich ist [Every year no more and no less wood must be harvested from the forests of the state than can sustainably be achieved by good management]." Just as in other sectors of society, forestry acquired economic indicators such as profitability much later – starting around the middle of the 19th century. The philosophy of sustainability was, however, never abandoned in professional forestry, and was sometimes used too rigidly.

In early history, the way to achieve sustainable management was almost exclusively to save the resource from exploitation. Later, however, forestry had to do what agriculture had done for centuries – increase production by improved management. It was also necessary to invest (by planting) for benefits in the distant future.

Even if most of the planning focused on wood production, it was soon recognized that other aspects benefited, such as wildlife and water. The difficulty was often that the various benefits of the same resource (the forest) were administered by different authorities, frequently with conflicting interests.

Awareness of the multitude of individual and common benefits offered by the forest was clearly spelled out in the literature (Grøn, 1931) and in governmental regulations in many European countries. In North America, government interest in the field can be found at least from about 1900 (Gregory, 1972). Up to World War II, sustainability in forestry was mainly a matter of the constant production of wood. Other aspects that were thought to suffer from intensive wood production (water, recreation) were in some areas protected by reducing or excluding wood production, but mostly such uses were regarded as safeguarded as long as deforestation was avoided.

This description of the early development of forestry mainly refers to what happened in the northern temperate zone. Forestry came to the tropics in quite different circumstances: it generally came later, to places with an abundance of wood and a very different ecology; and at first it came with limited responsibility for the welfare of the local society. Although the basis for the way of thinking in the following chapters is mainly Nordic, the topic is global and the problems are mostly picked from tropical forests because they are assumed to be the most challenging arena for forestry in the years to come. (For further reading, see Klose, 1985; Westoby, 1989.)

1.2.   Changes in society

Through the ages human society developed slowly. Sudden setbacks could come through catastrophes, wars, or epidemics (such as the Black Death), but the build-up would always take time. Development of human society is a complex process. It involves changes in cultural, social, and economic sectors based on increased knowledge about elements in science and technology. It is often related to changes in population with a corresponding need for increased production, trade, and communication. It is not always easy to say which part of the process comes first. What is more important is to secure balanced development among the sectors. Technology must fit actual economic development, production must meet the needs of the population, etc.

Of the many sectors of development let us scrutinize somewhat closer a few that are supposed to be of importance for management decisions in forestry. In doing so, it is not only past trends that are of interest. Because forest management today will influence possibilities a long way into the future, it is also necessary to see if future trends can be indicated.

1.2.1.   From few to many

World population is a very important factor in the formulation of resource policy in general, and clearly it is so also for forest resources.

In 1650, which was well before the industrial revolution, the world population has been estimated to have been about 0.5 billion, with a growth rate of some 0.3 percent per year (Meadows et al., 1972). From that time growth accelerated. The first billion was passed in about 1830, the second in 1930, and by 1950 the population had already reached 2.5 billion. In about 1987 the 5 billion mark was passed, with a growth rate of 1.6 percent per year. The growth rate is now decreasing, and according to UN forecasts it is assumed that the world population will stabilize towards the end of the 21st century. The number of people in the world is then estimated to be between 8 and 13 billion. Figures used by the FAO (1981) and WCED (1987a) indicate about 10 billion in the year 2100, as shown in Figure 1.

This means that in little more than a century – from 1987 to 2100 – the world's population will probably double, and more than 90 percent of the increase will come in the less developed parts of the world. The most rapid growth will occur in the first half of the century and will represent a challenge to mankind that has never been experienced before. It will be a challenge in many ways, and in particular to resource management. How can it be tackled so that present and future generations can meet their needs?

Even if the picture of future population shown in Figure 1 is less dramatic than Robert Malthus suggested in his essay "On the Principles of Population" (1798), it is still serious enough, and has been the background to much pessimistic thinking: How can the world produce enough food, or energy? How can we distribute products to where they are needed? People in many parts of the

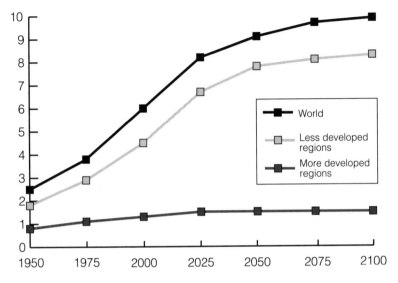

Figure 1. Past and prospective world population (billions)

world are already dying because they have insufficient supplies of food or water. This way of thinking is very understandable but, if it ends up in despair, it just make things worse. A realistic look at the future is necessary – not to scare people, but to discover where the most pressing problems are likely to be in the future, and what can be done now to reduce future pressure. The situation clearly challenges the meaning of the term "sustainability". This is taken up in Chapter 3.

1.2.2.   From subsistence to business

Economic systems have gradually evolved through history from subsistence to a monetary economy. It is now about 2600 years since money (coins) was first introduced, but in less developed regions a subsistence economy is still a way of life for large groups of people. Rural people in such regions live largely off their land, with some cash income from selling their produce or from paid work outside their farms. A noticeable part of this group has no cash income at all. These are all people who compare their own situation with that of other groups in society. Most of them do not wish to change their lifestyle drastically, but – wherever I have met them – they do want to reap more of the benefits that money can bring.

Locally we can still also find groups of people who have not yet reached the stage of agriculture. They live mainly from hunting and gathering, which require quite extensive areas and often a nomadic or semi-nomadic life. This can easily lead to conflicts in a society with a growing population and a developing economy. These conflicts have not always been solved in ways that accord with normal standards of human rights. In some of these societies, however, it seems to be a trend that the younger generation is looking for other ways of living, and even older people who for some years have had the opportunity to live more like the rest of society often do not want to return to a more or less nomadic life-style. Such natural developments towards a monetary economy are found, for example, in Sarawak, Malaysia. The general trend towards a monetary economy will most likely continue and lead to a situation where values in poor societies are to a greater extent measured in money. This will pave the way for more efficient economic development.

In the more developed parts of the world the increasing value of human labour has led to an organized market for labour, which is a further development of the economic system and results in a higher price of labour. This in turn has led to pressures for mechanization. An example can be taken from Norway: in the late 1930s it was possible to buy two man-days' work for the value of 1 m^3 pulpwood delivered roadside; 20 years later the value of 2 m^3 pulpwood was necessary to purchase one man-day's work. Without a process of mechanization very little industrial wood could have been harvested economically. The general result of the mechanization was that the value of the wood for the owner did not change very much and the number of forest workers decreased, but each of them increased their income to a level more comparable with that in the rest of society. A similar process of mechanization in a country where wages are low and unemployment high would not reduce the cost of logging, and it would send a larger part of the log value to the machine factories. Transfer of technology from a more developed to a less developed region should be done with great care, for environmental reasons as well. Heavy machinery is, however, often necessary to handle big logs.

Changes in economic systems and structures happen at different times and in different ways around the world. As a result, various groups of people are formed – even within one country – each with their own set of priorities regarding the value of the forest. The various groups and their priorities change over time. In richer societies – and particularly among people whose income derives from sources other than forest products – the non-monetary values of the forest are given higher priority.

Women in societies where subsistence farming is still a common way of life often have a close connection with the forest. It is their job to gather firewood and plants for food or medicine. In societies where the monetary economy prevails, women's influence on the forest rather comes from their role in decision-making on household consumption matters, and from their environmental interests. In the meantime, intimate knowledge about forests is reduced. These changes are in general different from what men are experiencing.

In relevant forums – such as the FAO – all aspects of the forest are discussed in meetings, and priorities are set. But priorities change: in the 1960s, production of industrial wood was high on the list; the 1970s brought rural development to the top of the agenda; and the 1980s was the decade when conservation dominated much of the discussion.

This situation can be confusing when forest management today influences which benefits can be reaped by both present and future generations. Each place and each time may have its own set of preferred benefits. Policy decisions will be necessary, and proper methods of forest management must be developed.

1.2.3.   From poor to rich

It is a general experience both for individuals and for nations that economic growth is hard to achieve, and hardest when you are poor. This does not necessarily mean that leaders in poor countries are less competent than their colleagues in less poor countries. It does mean, however, that poor societies have less "absorptive capacity" (educational level, infrastructure, etc.) to reap the benefits from the mainstream economies.

As an example we can take a look at the GDP (gross domestic product) per capita for the years 1970 and 1988 in 32 countries in Africa south of the Sahara (Norman, 1990). In Table 1 these countries are grouped according to the level of GDP per head in 1970. The annual growth in GDP between 1970 and 1988 is given in two ways, first as the average actual growth in US dollars and then as a percentage, assuming that the percentage growth rate was constant during the period. There are of course big differences between individual countries and between years in the period. For the purposes of this illustration, the averages in each group are used as they are without weighting them by the number of people in each country.

Table 1. Economic level and economic growth for 32 countries in Africa south of the Sahara

GDP/cap. in 1970 (USD)	GDP/cap. (USD)		Yearly increase	
	1970	1988	USD	%
< 100	60	206	8	7.1
100–200	145	456	17	6.6
200–300	255	915	37	7.4
> 300	460	1660	64	7.2

The example indicates that the initial phase of economic development – like most other pictures of growth (trees) – has a slow start and then accelerates. In this particular case the percentage growth rate seems to be relatively constant at the various levels of income. This means that the difference in income level between the groups in this phase is increasing. The difference between the top and bottom groups was USD 400 in 1970 and USD 1454 in 1988 even though both had almost exactly the same growth rate. Generally this is inevitable and does not mean that the rich are stealing from the poor. A similar situation can be found between groups of the population in a country.

In discussions about international development assistance emphasis is often placed on helping the poorest of the poor. Least developed countries such as those in the first group in the table will then be in focus. Inside each country the poorest section of the population will similarly be in focus. From an ethical point of view this must be right. The problem is how to achieve the intended result. If it is correct – as indicated above – that absorptive capacity is a major constraint on development, then it will be important to attack that capacity. This will normally not be most efficiently done by concentrating on the poorest people in a society. Always bearing in mind "the poorest of the poor", it is important to give the less poor a chance to lift the general absorptive capacity in a society.

The level of economic development tells us a great deal about the use of forest products in a country. This is in particular true for paper, which is increasingly an important forest product. I shall come back to quantities and available resources later (Chapter 3). Here I shall take a brief look at how economic growth under present conditions influences the use of wood.

GDP per capita is often used as a measure of the level of economic development, and there is a clear correlation between this indicator and the use of paper per capita. This appears even more clearly when GDP is recalculated as purchasing power parity (PPP), which corrects for the cost of living in the country concerned and expresses the result as a percentage of the figure for the USA. Figure 2 shows the relation between the consumption

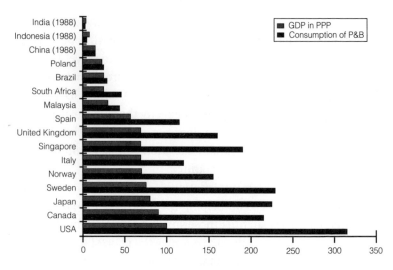

Figure 2. Paper consumption and purchasing power, 1990
Source: *The Economist* (1994), Wilkinson et al. (1992).

of paper and board (P&B) in kg per capita and GDP per capita in PPP.

In the period 1982–1990 world consumption of paper increased at a rate of 4.6 percent per year. In the same period the world population grew by 1.7 percent per year. The growth in paper consumption clearly has other causes than merely the increasing number of people. One important factor is likely to be improved economic standards. In this period the Pacific Rim of Asia experienced booming economic growth, and the consumption of paper in Asia as a whole grew by 8.4 percent per year. In Malaysia the figure was 10 percent. But even in highly developed countries the growth continues: the USA in the same period had a growth rate of 2.9 percent (Wilkinson et al., 1992).

The consumption of other wood products shows a different pattern. In the period 1975–1985 (FAO, 1985) world production of sawlogs and veneer logs grew by 1.6 percent per year, or at about the same rate as the population. Pulpwood and particles had annual growth of 1.8 percent. The reason for this figure being so much lower than that for paper and board (P&B) may partly be that the recycling of paper increased rapidly at that time. When the recycling has reached its maximum, the growth rate for pulpwood may increase.

Fuelwood and charcoal were estimated to account for more than 50 percent of all wood harvested in the world in 1985 (FAO, 1985). In a study for the World Commission on Environment and Development on "Energy 2000" (WCED, 1987b), wood was estimated to be the base for 35–42 percent of all renewable energy in the year 2000. This represents a greater share than hydro-power. The growth rate of fuelwood and charcoal was estimated at 2.5 percent in 1985 (FAO, 1985). Fuelwood is used all over the world (wherever there are trees), but it is in particular the source of energy for the poor regions. Scarcity is already an acute and worsening problem for large sections of the world's population.

1.2.4.   From standard of living to standard of life?

It has already been mentioned that the non-monetary values of forests are gaining more attention, particularly in the more developed parts of the world. This is an important development because many such values are likely to diminish if they are not taken account of in management. I shall not go into detail here – aspects of the problem will be taken up in later chapters. Here I will point out only that in many societies it is felt necessary to give priority to the economic and social standard of living before any surplus becomes available for securing non-monetary values, which are felt to mean more to other people. Much idealism is being invested in reducing the unnecessary use of natural resources. How far can we go towards an improved standard of life without destroying present and future generations' chances of meeting their own needs?

1.3.   Some milestones since World War II

1.3.1.   Organizations

The local and national level

The organizational and administrative pattern of forestry at the local and national level was mostly formed before World War II. This is particularly the case in the more developed countries, but even many of the less developed countries had made a good start.

National forest authorities (NFAs) varied a lot from country to country, depending largely on the importance of forests in the national economy and on the ownership structure. Two characteristic features seem to have been relatively common:

- Forestry was regarded as a minor sector in the national economy, too small to have its own ministry. Usually it was placed in the ministry of agriculture.
- The NFA was concerned with all aspects of managing the forest resource, including non-monetary values.

As a result of the accelerating changes in society described above, the post-war period led to certain modifications in the administrative pattern. Economic development in many cases led the NFA to concentrate on wood production. At the same time, other values increasingly came into focus, varying from one part of the world to another. Food production, watershed management, recreation, and conservation of nature were all fields where the forest had a role to play. These aspects of forestry were often allocated to sectors of the governmental administration other than the NFA. New knowledge in these fields and the NFA's necessary preoccupation with economic problems meant that the NFA would need additional expertise and more personnel to handle the situation. As some of these new administrative areas (e.g. recreation) also had aspects with no link to forestry, it could seem reasonable to establish new administrative units. As a consequence, however, in many cases *one* resource – the forest – is now handled by different authorities with conflicting interests. This can make administration very slow, and calls for very open-minded cooperation. A holistic view can easily be lost. If at the same time the NFA is part of a ministry where other interests are more in focus, this can make the situation even more difficult.

In the private sector, some *non-governmental organizations* (NGOs) were also in place before World War II. In countries with a high proportion of private ownership – as in northern Europe – the owners formed associations to deal with their common interests. Forestry workers had their unions, and in some countries idealistic associations were established to increase general knowledge about and interest in forestry. There were also organizations for recreation and for conservation of nature, but the conflict between their interests and those of wood-producing forestry was not very serious until intensified and mechanized forestry changed the forest and threatened to reduce non-monetary values. In large parts of the world this characterizes the post-war period, which was the time when environmental NGOs started influencing forestry.

The international level

Cooperation in research had started as early as 1892 with the founding of the International Union of Forest Research Organizations (IUFRO). In addition, a World Forestry Congress has been regularly convened since the 1920s. Similar organized cooperation can be found quite early in some regions. These were all efforts to promote knowledge and understanding of what the forest could mean to mankind if properly managed.

The first *intergovernmental attempt* to establish organized cooperation was in 1932 when the Comité International du Bois (CIB) was set up, sponsored by the League of Nations (Glesinger, 1949).

This organization had a rather short life.

World War II led to the establishment of a new organization – the United Nations (UN). At its first gathering in May 1943 at Hot Springs, the UN agency for food and agriculture, the FAO, took up the idea of serving forestry as well as agriculture. What was left of the CIB was a few years later taken over by the FAO, which in the post-war period has been the main intergovernmental organization for forestry. Other UN agencies too were successively involved, such as the United Nations Development Programme (UNDP), the World Bank, and, later, the United Nations Environmental Programme (UNEP). Through the collection and dissemination of statistics and other information, through multinational meetings and seminars, and by channelling technical and financial assistance in particular to less developed countries, the FAO and the other organizations established a base for global thinking in forestry. This work has recently reached an important stage with the publication of *The UN–ECE/FAO 1990 Forest Resources Assessment* (ECE/FAO, 1993a,b).

Based on the UN declaration and programme on the establishment of a New International Economic Order, and as a result of resolutions and discussions in the United Nations Conference on Trade and Development (UNCTAD), a new organization came into being. The International Tropical Timber Organization (ITTO) has been active since 1985. It is not a UN agency but an intergovernmental organization building on an Agreement signed by member countries in 1983. The Agreement defines objectives, the first of which is: "To provide an effective framework for co-operation and consultation between tropical timber producing and consuming members with regard to all relevant aspects of the tropical timber economy." This organization covers the trade aspect more actively than the FAO can do, but is on the other hand limited to dealing with only tropical timber.

In the last quarter of the century two important new institutions in the field of forestry research have been established: the International Council for Research on Agro-Forestry (ICRAF) and the Center for International Forestry Research (CIFOR).

Before we leave the intergovernmental field, one more establishment should be mentioned. The Tropical Forestry Action Programme (TFAP) is not an organization but an unofficial programme through which various agencies and organizations in the field try to stimulate and coordinate relevant activities, as well as make them harmonious parts of more general plans for development. The TFAP had its first meeting in 1985 and for the time being has a small secretariat in the FAO. An independent high-level review of the TFAP was initiated in 1990 by the Director General of the FAO. To quote from the review report:

> The response to the TFAP exceeded expectations at a time when FAO was ill-prepared to meet the demands made on it by member governments. However, despite some criticism, misinterpretations and misgivings, no comparable alternative global initiative has emerged. We believe that the TFAP should be continued, although with a changed mandate and modifications as suggested later in this Report.

The map of the intergovernmental organization of forestry has changed from a single organization (the FAO) just after the war with a single focus – wood production – to several organizations and agencies, all trying to cover a multitude of products and benefits. The old way of building organizations where forestry is included as part of something else that does not cover the whole field may not be the best way any more. A more up-to-date background to rethinking the organizational pattern of international forestry can be found in the agreement adopted by the United Nations Conference on Environment and Development (UNCED) in Rio de Janeiro in 1992. The first part of the agreement reads as follows:

> Non-legally binding authoritative statement of principles for a global consensus on the management, conservation and sustainable development of all types of forests:

Preamble

(a) The subject of forests is related to the entire range of environmental and development issues and opportunities, including the right to socio-economic development on a sustainable basis.

(b) The guiding objective of these principles is to contribute to the management, conservation and sustainable development of forests and to provide for their multiple and complementary functions and uses.

(c) Forestry issues and opportunities should be examined in a holistic and balanced manner within the overall context of environment and development, taking into consideration the multiple functions and uses of forests, including traditional uses, and the likely economic and social stress when these uses are constrained or restricted, as well as the potential for development that sustainable forest management can offer.

(d) These principles reflect a first global consensus on forests. In committing themselves to the prompt implementation of these principles, countries also decide to keep them under assessment for their adequacy with regard to further international co-operation on forest issues.

(e) These principles should apply to all types of forests, both natural and planted, in all geographic regions and climatic zones, including austral, boreal, sub temperate, temperate, subtropical and tropical.

(f) All types of forests embody complex and unique ecological processes which are the basis for their present and potential capacity to provide resources to satisfy human needs as well as environmental values, and as such their sound management and conservation is of concern to the Governments of the countries to which they belong and are of value to local communities and to the environment as a whole.

(g) Forests are essential to economic development and the maintenance of all forms of life.

(h) Recognising that the responsibility for forest management, conservation and sustainable development is in many States allocated among federal/national, state/provincial and local levels of government, each State, in accordance with its constitution and/or national legislation, should pursue these principles at the appropriate level of government.

(UNCED, 1992b:1)

The continued work to find an organizational pattern that can secure holistic and balanced management of the forest resource is most important. It will determine to what degree the forests of the world can support sustainable development for humanity. An Ad-Hoc Intergovernmental Panel has been established to discuss problems in this field.

International non-governmental organizations (INGOs) have increasingly played a role in world forestry in the post-war period. We find them in industries (for example, Asia-Pacific Forest Industries) and in trade (for example, Union pour le Commerce des Bois Tropicaux dans la CEE – UCBT). More impact on forest management has, however, come from the INGOs on the environmental side. Discussions between economic and non-monetary interests, between sector interests and global views, and between consumption and conservation arguments often have limited effect, whereas positive action to assist in solving problems has achieved good results. There is a multitude of environmental INGOs. It is beyond the scope of this book to describe them. Three examples that are known to have contributed to improvements in forest management should be mentioned here:

- the International Union for the Conservation of Nature and Natural Resources (IUCN), which was founded in 1948 and has both governments and national NGOs as members.
- the World Wide Fund for Nature (WWF), which was founded in 1961 and works to generate moral and financial support for safeguarding the living world.
- the International Institute for Environment and Development (IIED), which in latter years has taken part in very useful analytical work in forestry.

Public awareness and understanding of the possibilities and limitations in nature and its relation to human needs are essential for the smooth and flexible management of the forest resource towards a balanced optimum. INGOs can do a great job here if they can find the right balance.

1.3.2.   Literature

One way of tracing the growing understanding and concern for forestry problems during the post-war period is of course to follow the trends in the relevant literature. Out of the overwhelming multitude of books, reports, and articles in the field, I shall attempt to pick a few and very briefly indicate the turbulent landscape of ideas that professional forestry has had to find its way through.

Egon Glesinger's book *The Coming Age of Wood* was published in 1949. Glesinger had been Secretary General of the CIB, but was at this time with the FAO. A quotation from his book may explain his message:

> Wood will become the characteristic raw material of our civilization because it has three attributes which make it unique among all raw materials.
> 1. Wood is universal. Wood has become a raw material that can satisfy almost every requirement of existence . . .
> 2. Wood is abundant . . . And only a fraction of the world's forest resources is now utilised.
> 3. Wood is inexhaustible. The forest is not a mine that eventually will be depleted, but a cropland. *Provided* trees are harvested as a crop and the forest is sustained by proper management, wood will forever yield all the material the human race can conceivably require.

In the early post-war period this was a message with considerable appeal.

Rachel Carson's *Silent Spring* came out in 1962. It was mainly a dramatic warning against the contamination of nature with herbicides and insecticides. In this and also in more general terms she has views that can be seen as critical of intensive forestry (page 64):

> The earth's vegetation is part of a web of life in which there are intimate and essential relations between plants and the earth, between plants and other plants, between plants and animals. Sometimes we have no choice but to disturb these relationships, but we should do so thoughtfully, with full awareness that what we do may have consequences remote in time and place.

This was a reminder to foresters that in an era of technology and economy, old lessons about biology should not be forgotten, and new lessons had to be learned.

"The Club of Rome" initiated a research programme led by Dennis Meadows and involving a group of researchers covering a number of relevant fields. The main idea was that the problems encountered in studying future development were so complex and interdependent that traditional institutions and political guidelines could not handle them. Meadows and his team produced a book, *The Limits to Growth*, in 1972. It took up problems concerning food production, pollution, and the use of non-renewable resources. Further analysis and discussions were undertaken 20 years later in *Beyond the Limits* (Meadows et al., 1991). In particular, the message that some important resources (minerals, coal, oil) not only are non-renewable but probably would be scarce – economically and even technically – within a century or two was picked up by the forestry community. Will wood in the future – as a renewable resource – have to substitute for important non-renewable materials? Could we produce that much?

In 1980 the Council of Environmental Quality and the Department of State in the USA came up with *The Global 2000 Report to the President*. The forestry part of the report is rather pessimistic and, looking ahead to the year 2000, it says: "Growing stock per capita is expected to decline 47 percent world-wide and 63 percent in Less Developed Countries" (page 26).

Also in 1980 came the *World Conservation Strategy* prepared by the IUCN, WWF, and UNEP in collaboration with the FAO and Unesco. The following summary is of great interest to forestry (section 7):

> The requirements for achieving the objectives of conservation can be summarized as follows:
> - the maintenance of essential ecological processes and life-support systems primarily requires rational planning and allocation of uses and high quality of those uses;

- the preservation of genetic diversity primarily requires the timely collection of genetic material and its protection in banks, plantations and so on, in the case of off site preservation; and ecosystem protection in the case of on site preservation;
- the sustainable utilisation of ecosystems and species requires knowledge of the productive capacities of those resources and measures to ensure that utilization does not exceed those capacities.

Our Common Future – the report from the World Commission on Environment and Development (WCED – also called the Brundtland Commission) came in 1987. It lifted problems of the type mentioned above to a level that made them visible both to decision-makers and to the general public. It also outlined the interaction between resource management and general human development. It has a message for all types of resource management, including forestry. Its conclusions in the chapter "Towards Sustainable Development" are important guidelines to the actual situation:

In its broadest sense, the strategy for sustainable development aims to promote harmony among human beings and between humanity and nature. In the specific context of the development and environment crises of the 1980's, which current national and international political and economical institutions have not and perhaps cannot overcome, the pursuit of sustainable development requires:
- a political system that secures effective citizen participation in decision making,
- an economic system that is able to generate surpluses and technical knowledge on a self-reliant and sustained basis,
- a social system that provides for solutions for the tensions arising from disharmonious development,
- a production system that respects the obligation to preserve the ecological base for development,
- a technological system that can search continuously for new solutions,
- an international system that fosters sustainable patterns of trade and finance, and
- an administrative system that is flexible and has the capacity for self-correction.

These requirements are more in the nature of goals that should underlie national and international action on development. What matters is the sincerity with which these goals are pursued and the effectiveness with which departures from them are corrected.

(WCED, 1987a:65)

A number of useful publications have in recent decades come from the FAO. Only two will be mentioned here. *The Challenge of Sustainable Forest Management* was published in 1993 and is a brief and relatively popular presentation of actual forestry problems. To quote the Assistant Director-General in his preface: "This book is a contribution not only to increasing public awareness of the issues involved but also eventually to the implementation of sustainable forest management and of sustainable land use."

Together with the United Nations Economic Commission for Europe (ECE), the FAO published *The UN–ECE/FAO 1990 Forest Resources Assessment* (ECE/FAO, 1993a,b), with the *Global Synthesis* following in 1995 (FAO, 1995). This comprehensive global overview of forest resources and their uses is based on the best available statistics and estimates. It is a major effort, and will be a useful background to future discussions about resource management. Some results of the assessment will be given in Chapter 2.

1.4. The confusion of today

The changes in society mentioned above have in many cases led to changed attitudes to what sort of forest management would be beneficial. Depending on personal interests and available information, the priority accorded to the various benefits will vary. As population grows and interests diversify, it is increasingly difficult to satisfy everyone.

Organizational development and trends in the literature indicate a growing public recognition of the various benefits offered by forests. This is coupled with the fear – or even the conviction – that the actual management of the forest resource does not secure those benefits.

An increasing number of people will feel that something is wrong with forestry. They may be right. The problem is that their interests, information, and therefore also their views are different. Various groups of people – whether organized or not – will have difficulty reaching agreement on "what is wrong". They might, however, each from their own outlook, more easily agree that "something is wrong" or, to be a bit more specific, that there is a "suboptimal aggregate benefit of forests". This situation has led to some confusion both about what we as a society want and about how we should organize our work.

An attempt to structure the "wrongs" has been made in Figure 3 in a form that has been called a "problem tree". Such a "tree" was originally formed (ITTO, 1992) to discuss local problems and establish a relevant project to solve them. As it is used here, it has a more general background. The figure indicates in very broad terms a number of "wrongs" that serve as roots in the problem tree, whose top is the suboptimal benefit.

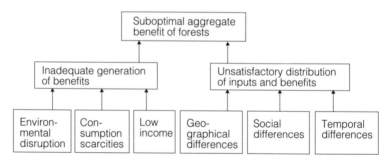

Figure 3. The problem tree

The bottom line – the roots of this problem tree, or the roots of dissatisfaction – consists of the undesired managerial conse-quences that often occur. Chapter 3 will briefly attempt to uncover some of the main roots, and Chapters 4, 5, and 6 will discuss some possible ways of increasing the aggregate benefit.

THE WORLD'S FOREST – main vegetation zones

Cool coniferous forest

Temperate mixed forest

Warm temperate moist forest

Tropical moist evergreen forest

Tropical moist deciduous forest

Dry forest

Chapter 2

World forest resources – a brief overview

A discussion of sustainable and multipurpose management of the world's forests needs a comprehensive and reliable description of the resources, their use, their potential, and their actual development. The best available background is found in *The UN–ECE/FAO 1990 Forest Resources Assessment* (ECE/FAO, 1993a,b) and the *Global Synthesis* (FAO, 1995). The brief overview in this chapter is mainly taken from these publications. A more detailed study would necessitate a closer look at the assessment itself and at other literature concerning consumption, environmental problems, etc.

Data of the kind presented in the assessment must be regarded as approximations. The base information is incomplete, and inventory procedures, classifications, and definitions are not quite the same everywhere. For the developing countries, changes in forest cover have been estimated using a model with population pressure and ecological zone as variables (in Africa, parallel interpretation of new and historical satellite images has been used). For developed countries, information has been collected by means of a questionnaire. With this in mind, I shall here take a brief look at the global situation of the forests. This overview is considered sufficient for the purposes of this book.

← The World's Forest. Main Vegetation Zones. © UNASYLVA, FAO, 1976.

2.1.   Ecological zones

Cool coniferous forest (including the Taiga) is found in a broad belt bordering the northern Arctic region. It covers most of Canada and Alaska, and stretches through Scandinavia and northern Russia all the way to Kamchatka. Coniferous trees are dominant, but several broad-leaved species also find optimal conditions in this zone.

Temperate mixed forest is the next belt, covering much of the USA, continental Europe, and parts of Russia and China. The warmer climate is optimal for additional species of trees and plants, although the two zones have many species in common. Broad-leaved trees will, however, be more dominant in the natural vegetation in this zone. In the southern hemisphere we find similar vegetation in parts of Chile and Argentina.

Warm temperate moist forest is found, for example, in the southeastern USA and in the coastal part of southern China. It is still mainly in the temperate zone, but has much in common with certain types of tropical forest.

For the tropics, the FAO assessment distinguishes between a number of Lowland formations, an Upland formation, and a Non-forest zone. Broad-leaved trees are generally dominant in the tropics. Of the 1756 million hectares (ha) of tropical forest, the greater part – 1544 million ha – is in the Lowland formation. Forest makes up 44 percent of the land area in the Lowland formation. The forest in the Lowland formation is again divided into three zones:

(1) *Wet/very moist forest* with rainfall of over 2000 mm/year. Here we find the real rain forest with evergreen broad-leaved trees and very high biodiversity. It is estimated to cover an area of 718 million ha or about 40 percent of the tropical forest. In the wet zone 77 percent of the land area is still covered by forest.

(2) *Moist deciduous forest* with rainfall of 1000–2000 mm/year. Most trees are without leaves for a certain period each year. The area is estimated to be 587 million ha.

(3) *Dry/very dry forest* with rainfall of less than 1000 mm/year. This accounts for about 238 million ha in the tropics. There are, however, similar types of forest in the temperate zone, for example in Mexico and Australia.

One important difference between the zones is the species richness – the biodiversity. The total number of higher plants (including ferns, cycads, conifers, and broad-leaved species) in the world is estimated to be some 270 000. Whereas about 85 000 of these plants can be found in Latin America, only 17 000 are found in North America (FAO, 1995). This is in spite of North America having a larger land area and fewer people per km² than Latin America. Differences in land use are certainly part of the reason, but it does illustrate a general trend in biodiversity. No other ecological zone is so rich in species as the tropical rain forest.

2.2.   Land use

Decisions concerning land use are generally a national prerogative. They will often depend on traditional rights, pressure from interest groups, and economic strength.

The more developed (industrialized) countries with noteworthy forest resources are largely in the temperate and boreal zones. Some less developed and developing countries with forest resources are found in the temperate zone, but of the forest area in all less developed countries about 87 percent is tropical. In this book, land use is seen more in relation to the degree of development than to ecological zone. This will, however, at the same time give an approximate description of the relation between the two main zones – the temperate and the tropical.

The land use of the countries included in the 1990 assessment is shown in Table 2.

Nordic forest. © sebra-film co.

Tropical rain forest. © sebra-film co.

Table 2. World land use (million ha)

Agricultural land		1 442
Other non-wooded land		6 374
Forest and other wooded land (FOWL)		5 120
of which: Other wooded land	1 678	
Forest	3 442	
Total land area		12 936

It is indicated that there is the potential to expand the agricultural area to more than 3000 million ha. This potential is mainly found in the less developed countries, where it is most needed. "Other non-wooded land" will generally have low productivity: about 800 million ha are extreme desert; other areas are arid or shallow; some have too much water, and some have permafrost. Urban developments are spreading, often on good soil. It therefore seems inevitable that considerable forest areas in the future will have to be used for agriculture. "Other wooded land" does produce some wood for human consumption, and this production could be greatly improved in parts of the area. In addition, trees outside the forest yield some wood products.

Disaggregating some of the above figures, we find important differences between regions in the world (Table 3).

Table 3. Forest and other wooded land, developed and less developed countries

	FOWL m.ha	Forest	
		m.ha	ha/cap.
More developed countries	2064	1432	1.1
Less developed countries	3056	2010	0.5
World	5120	3442	0.6

It is worth noting that the less developed countries have less than half the area of forest per capita compared with the more developed countries.

Of particular interest here is the *forest* area. For a closer look at how this area is used, we must look at the estimations for the economic regions separately.

The more developed countries

Total forest area		1432 m.ha
of which: exploitable part	898	
unexploitable part	534	

The unexploitable forest is of two types:

(1) Forest where physical productivity is too low or harvesting and transportation costs are too high to warrant commercial wood harvesting. This comprises about two-thirds of the area.

(2) Forest with strict legal restrictions on wood production, e.g. national parks, nature reserves, and other protected areas such as those of special scientific, historical, or cultural interest. This comprises about one-third of the area or about 11 percent of the total forest area.

The less developed countries

A similar classification of these countries is not yet possible. A limited part of the *tropical forest* is, however, notified as reserved and classified by forest function. This will give an impression of the situation:

Total tropical forest		1756 m.ha
Notified as reserved		522
of which reserved for:		
Wood production	309	
Protection	142	
Conservation	71	

Tropical plantation. © sebra-film co.

Protection means here that the function is to conserve soil and water, while conservation refers to biological conservation.

A special study (ECE/FAO, 1993b) in the more developed countries shows that increasing emphasis is given to policy and planning activities relating to protection, nature conservation, and recreation. Other functions – such as wood production – seem to have a relatively stable emphasis.

2.3. Some important trends

The findings in the 1990 Forest Resources Assessment allow significant conclusions to be drawn regarding the development of the world's forests in the period 1980–1990.

2.3.1. Forest area

In the total region of more developed countries the forest area (1432 million ha) has changed very little. Areas that have been taken for other uses are of the same magnitude as areas of afforestation and natural extension of the forest.

In the less developed region there has been an annual deforestation of 16.3 million ha, of which 15.4 million ha are in the tropics. The annual increase in forest area through forest plantation was 4.1 million ha. The net decrease in forest area has therefore been about 12.2 million ha per year. Of this area about 2.0 million ha become "other wooded land" – mainly through shifting agriculture.

Taking the annual tropical deforestation of 15.4 million ha, this is equivalent to 0.8 percent per year. The corresponding figures for the three Lowland formations are as follows: rain forest – 0.6 percent; moist deciduous – 1.0 percent; dry – 0.9 percent. These differences probably reflect the pressure for agricultural land and suitability for cultivation.

Area changes other than deforestation and plantation are also important to note. Fragmentation of the forest area is an ongoing process as agriculture successively takes over. A good part of what was fragmented forest in 1980 had become non-wooded land in 1990, and a similar proportion of the 1980 forest was being fragmented during the period up to 1990.

2.3.2.   Volume and biomass

Although expansion of agriculture seems to be the main reason for deforestation, wood harvesting for industry or local consumption may be more important for the reduction in volume in areas where the forest remains.

The growing stock – the volume above ground of wood (including bark) with a minimum diameter of 10 cm – in the forests of the world is estimated to be 384 billion m^3. This corresponds to 114 m^3/ha, and this figure is approximately the same for the two economic regions. The biomass in tons/ha is, however, considerably higher in the less developed region than in the more developed region – 169 and 79 tons/ha, respectively. This is mainly the result of different species having different forms and wood quality, but weaknesses of the assessment may also play a role.

In the more developed regions, it is clear that fellings (the volume of trees felled) are significantly less than the net annual increment (NAI) in exploitable forest; they are estimated to be about 74 percent of the NAI. This means that during the 1980s there was increasing wood volume and biomass. Indications are that the annual increase in volume during this period was about 0.6 percent.

In the less developed regions, similar estimates are not available. However, it is apparent from the changes in area description within the forest that degradation is occurring from dense forest to open forest, which means reduced wood volume and biomass. The annual loss of biomass due to deforestation is estimated to be 0.9 percent. When the loss due to degradation is added to this figure, *the unavoidable assumption must be that the volume and biomass of the world's forests decreased during the 1980s.*

It is important to observe that the forest resource in the more developed part of the world has increased whereas in the less developed part it has decreased. This means that reduction of biomass is mainly occurring in poor countries, most of which are located in the tropics. These are the countries that have about 60 percent of the world's forest area, but harvest only a quarter of the world's industrial wood (1992). They do, however, harvest 87 percent of the wood used for energy. This is particularly depleting for the forest where it is sparse for natural reasons (arid and shallow areas, etc.). The less developed countries also in general have high population density and population growth, which necessitate intensified and/or expanding agriculture. This set of characteristics makes it understandable that what is happening in these developing countries is most often the result of socio-economic pressure beyond the natural capacity of the resources. It has to do with the struggle for survival.

The region that is now regarded as more developed went through a similar harvesting of resources with low productivity in the early stages of industrialization.

2.3.3.   Biodiversity

As regards the tropical forest, the 1990 assessment took up the question of how biodiversity is influenced by deforestation. Only higher plants are included in the study. It shows that wet areas (rain forest) have higher species richness than moist areas, and that dry areas have the poorest biodiversity. Within each of these ecological zones, large areas have more species than small areas. Upland formations are also included in the study.

The effect of normal harvesting or degradation was not considered in the study. It is, however, an important question, which will need attention in more local studies.

Chapter 3

Undesired and challenging managerial consequences

At the end of Chapter 1, I asked, "what is wrong?", and in Figure 3 some indications were given of the broad areas in which the roots of dissatisfaction could lie. These are all areas that have to be taken into account when managing the forest resource. Westoby (1987:192) is focusing on problems of this kind when he says:

> The challenge lies in the understanding we have achieved of the potential contribution which forestry can make to development and rising welfare. For, contrary to what many outsiders believe, forestry is not, in its essence, about trees. It is about people. It is about trees only so far as they can serve the needs of people.

Applied to the presentation given here, this means that *forestry can be defined as the activity concerned with establishing, managing and utilising forest ecosystems aiming at satisfying human needs.*

In this chapter I will take a somewhat closer look at the roots of dissatisfaction as they are presented in Figure 3. Each area is of course big enough to fill not only a book but a good part of a library. This presentation does not therefore pretend to cover all aspects. The point here is to give an idea of the multiplicity of aspects that in varying constellations can be the basis for decisions in forest management. A broader discussion of many of the issues in this chapter can be found in Sharma et al. (1992).

Industrial logging. © NPS/K.&K. Ammann

3.1.   Environmental disruption

Disruption or damage to the environment can occur as the result of human interference. It can also occur owing to natural catastrophes (fire, wind, flooding, avalanches), which could sometimes have been prevented by human interference. Some damage has an impact mainly on the forest itself (internal damage). Other damage creates problems for areas outside the forest (external damage).

3.1.1.   Internal damage

Deforestation

Deforestation is here defined as change of land use from forest to other land use or depletion of forest crown cover to less than 10 percent (see FAO, 1993).

In the temperate zone, clearcutting is the normal way to establish regeneration. It is then a natural part of management and has nothing to do with deforestation. In natural forest in the tropical zone, the climate, the number of species, and even market conditions make it natural to harvest only big trees and leave the smaller ones for further production. This is also in most places regulated by law. Logging in itself is therefore not a major cause of deforestation, but roads that have been built for logging are frequently used as the base for agricultural expansion. Although shifting agriculture is the main cause of deforestation in the tropics, organized development of permanent agriculture, dams, roads, and necessary urban expansion also need areas from the forest. The annual rate of deforestation in the tropics for the period 1981–1990 has been estimated at 15.4 million hectares (ha) by the ECE/FAO (1993c), which is about 0.8 percent of the forest area. About 41 percent of the tropical forest area is rain forest with an annual deforestation rate of 4.6 million ha, or 0.6 percent (see section 2.3.1).

Degradation

> Degradation, as distinct to deforestation, means that the production potential of the forest area is reduced through some outside, mostly human, factors. Degradation can include damage both to the soil and to the growing stock.
>
> (Bostrøm, 1986:2)

Degradation can have a multitude of causes. Logging with heavy machinery can damage trees that should be left for further production. It can also cause erosion on vulnerable soil. In districts where the population is high and forest is scarce, the local use of fuelwood can lead to degradation of the growing stock. The same can happen as a result of grazing. Again natural disasters, pests, and diseases can play a role. When degradation is severe it can result in deforestation. The annual rate of degradation of tropical forest in the period 1981–1990 has been estimated at about 0.2 percent by the ECE/FAO (1993c).

Fragmentation

In areas of relatively dense population, instead of a large continuous forest area the closed forest will often be split up into "islands" in a "sea" of other land. In such situations both degradation and deforestation occur frequently.

Changes in biodiversity

Even in untouched forest, changes in biodiversity can occur and may be enhanced by, for instance, pollution, which can travel a long way through the atmosphere. From a conservation point of view, changes in biodiversity can be seen as a type of degradation. This is particularly the case when endangered species disappear. The IUCN's guidelines, *Conserving Biological Diversity in Managed Tropical Forests*, say in the first of the basic principles:

Shifting agriculture, Thailand. © sebra-film co.

Flood in rainy season, Thailand. © sebra-film co.

Erosion, Tanzania. © sebra-film co.

Any disturbance of a forest, natural or man-induced, will alter it as a habitat for animal or plant species. Small-scale disturbances may enhance structural, floristic and faunistic diversity. Large-scale disturbances tend to simplify the ecosystem and result in loss of biological diversity.

(Blockhus et al., 1992:6; see also section 2.3.3)

Some aspects of local changes in biodiversity can be illustrated by research undertaken in tropical rain forest in Sarawak, Malaysia. Primack and Hall (1992) have been studying changes in the number of tree species over a period of 20 years in untouched primary forest. In their plots of 0.6 ha each they found that 65–70 percent of the species were present both in 1965 and in 1985. Species present in 1965 but absent in 1985 accounted for approximately 12 percent of the total number of species of each site. Immigration of new species onto the plots (present as trees in 1985 but not present in 1965) accounted for 16–24 percent of species recorded. These changes in untouched primary forest (on relatively small plots) seem to be substantial even within a period of only 20 years. For larger areas the number of species will be more stable.

The effect of logging on botanical diversity was not discussed by Primack and Hall. Zoological diversity, however, was studied – also in Sarawak – by Bennet and Dahaban. Wildlife populations were monitored continuously from one year before to one year after logging. They report that: "One year after logging, the total number of animal species had declined but not significantly. The composition of species did change somewhat, however" (1992).

Bennet and Dahaban also studied the wildlife in secondary forest that had been cultivated some 30 years earlier. "The number of species of diurnal primates, squirrels, all mammals and all birds were slightly higher in the two old cultivation (temuda) areas than in primary forest" (1992). There were, however, certain changes in the composition, dependent on their distance from primary forest.

The studies quoted above were concerned with the number of species. The tendency in discussions of biodiversity now seems to be to focus more on ecological functions than on the number of species. This would be more useful for our purposes. However, too little seems to be known about the effect on biodiversity – whichever definition is used – of various kinds of forest management, but the statement from the IUCN guidelines quoted above seems to be a realistic assumption.

Taking the global view through geological history, Gore (1989:669) says: "Extinction has claimed 99 percent of all species that have ever lived – many of them victims of 'background' extinction, the piecemeal disappearance of species due to small-scale environmental changes. Others perished in one of earth's major mass extinctions detected in the fossil record."

Aesthetic character

The forest landscape can be influenced by what happens in the forest, and changes will often be seen as depletion or degradation. Such evaluations will include subjective factors, but people living under similar conditions, in the same area, will tend to have similar views.

Use of the forest for one particular benefit can often be regarded as abuse – or even depletion – by those who have a preference for other benefits. Some comments on the relationship between benefits will be given in the next chapter (see section 4.1.2).

3.1.2.   External damage

Through its influence on soil, water, and atmosphere, forest management may lead to damage outside the forest itself.

Water

Water is always causing erosion. That is how the fertile lowlands around rivers have been built up (the Nile, the Ganges). Deforestation or severe degradation will cause more water to run straight off in rainy periods. This often leads to increased erosion and landslides, to disastrous flooding in downstream areas, and to lack of adequate water supplies in dry periods.

As regards dry areas, an interesting line of thinking goes as follows. In some places (e.g. West Africa) the increased run-off in coastal areas caused by deforestation reduces the amount of water stored in the ground and the water released back into the atmosphere by vegetation. This in turn reduces the humidity in the air that is carried by the prevailing wind towards areas of expanding desert. Deforestation in some coastal areas is therefore one of the reasons for the ongoing process of desertification in the interior. This theory sounds quite plausible but may be difficult to prove experimentally.

Soil

A lack of vegetation can also cause wind erosion. Sand dunes are often shifted, again with desertification as a result. Existing forest in such areas must be managed with great care, and, if the forest has already gone, afforestation may sometimes be a useful tool in protecting adjacent areas.

Atmosphere

One impact of forest management that does not limit itself to neighbouring districts and countries but is a global concern is the effect on the chemical composition of the atmosphere. In the discussion about possible "global warming" in particular, atmospheric carbon dioxide has been shown to be important. According to the FAO (1993):

> The main contribution to the increasing concentration of greenhouse gases in the atmosphere has been from the burning of fossil fuels. This is likely to remain the case over the next few decades. The Intergovernmental Panel on Climate Change (IPCC) projected that fossil fuel

consumption would contribute 65 percent of the greenhouse effect from 1990 to 2025. The contribution from deforestation and biomass burning over the same period is projected to be 15 percent.

The significance of the forest in this connection lies in the fact that about half of the dry matter in wood is carbon. If, instead of reducing the amount of wood (alive or dead) in the world, we can increase it, the effect on carbon dioxide (CO_2) in the atmosphere should have a similarly positive effect on the climatic situation. The total biomass (stems, branches, stump, and roots) necessary for the production of 1 m^3 of industrial wood represents about 1.5 metric tons of CO_2, which is equivalent to the emissions from some 640 litres of gasoline.

The increase in wood should not necessarily be achieved by reduced harvesting. It could be done more efficiently by expanding the forest area (plantations) and by forest management that builds up forests with a greater volume of wood per hectare. This has been done in several countries in the temperate zone. Developments in Norway, for example, have been as follows:

> Volume of growing stock has shown a considerable increase since the start of the survey in the 1920s. The volume of trees has grown from approximately 300 million cubic meters to between 500 and 600 million cubic meters (excluding bark). At the same time, the annual increment has increased from 10–11 million cubic meters to approximately 19 million cubic meters. The number of trees has shown a remarkable trend to multiply, especially for the larger diameter classes.
>
> (Tomter, 1993)

In the same period the annual harvest of wood for sale has increased by some 75 percent. The harvest for home use has probably gone down a bit, but not by enough to outweigh the increase in wood for sale. The harvest has, however, always been well below the increment.

Tropical rain forests are often mentioned in discussions about forests as a carbon sink. It is important to note that all wood – everywhere – locks up carbon that might otherwise be in the atmosphere. The special situation in untouched tropical moist forest is that (just like other untouched forests) it does not absorb or produce anything. Living trees absorb a certain amount of carbon from the atmosphere. Dead rotting trees give back about the same amount. But, like all other wood, the wood in the rain forest represents a stock of carbon that is not released into the atmosphere. If we could build up this carbon sink in a global programme – and keep it high – we would contribute to reducing the risk created by the burning of fossil fuel.

3.2.   Consumption scarcities

Forestry grew to be a profession and a science over the past two or three centuries basically because local scarcity of wood – or the fear of future scarcity – became a reality. In the 20th century, with its rapidly growing population and increased use of natural resources, there has been increasing concern about the scarcity of various resources.

A resource is here understood as scarce when it is unable to fulfil its role in the process of sustainable development. "Sustainable development is development that meets the needs of the present without compromising the ability of future generations to meet their own needs" (WCED, 1987a).

3.2.1.   Types of natural resources

For the discussion here it is practical to distinguish between three types of natural resources – finite, re-usable, and renewable – because a possible scarcity in any of them will need different kinds of remedial action. Figure 4 illustrates how different types of resources involve different management options in order to serve sustainable development.

		Possible management		
		Save	Maintain	Produce
Resource type	Finite	X		
	Re-usable	X	X	
	Renewable	X	X	X

Figure 4. Possible remedial action for scarcity

Finite or non-renewable resources

Finite or non-renewable natural resources exist only in given quantities. When they have been used up they cannot be renewed or used again for a similar purpose. Petroleum is an example of this type of resource. The way to avoid scarcity of a finite resource must be to *limit consumption*, either by better utilization or by finding alternative resources.

Re-usable resources

Re-usable resources can be used again and again if they are properly exploited. They may be improved or destroyed to varying degrees.

Productive land is one such resource. To prevent the deterioration of productive land, suitable forestry activities are useful tools – and sometimes the only practical ones. If the deterioration of productive land becomes extreme (erosion, desertification) it may be very difficult – or impossible – to bring the land back to a productive status. On the other hand, productive land can be improved to increase its biological productivity (by drainage, irrigation, fertilization) or its economic potential (by road building).

Water is another re-usable resource. To be useful – and not destructive – the quantity and quality of the water in the soil and in streams need to be regulated. In this connection, too, forestry activities are essential.

The genetic resources in plants and animals are also re-usable. In the forest most of them are easily maintained by careful management. However, some are to a greater or lesser extent rare and endangered. In the general course of biological development, some species disappear and new ones appear without human interference, but human activity has a clear tendency to speed up the disappearance of vulnerable species.

The normal way to avoid scarcity of re-usable resources is to actively *maintain* them. Endangered species will need special attention to save them. This is mostly done *in situ* (in their natural environment), but can also to some extent be done *ex situ* (in a specially designed environment). In either case the process involves prioritization based on principles linked to the problem of sustainable development.

Renewable resources

In the forest, renewable natural resources are the living plants and animals, or products derived from them. Trees will normally renew themselves as long as the re-usable resources (land, water, genes) that are their necessary basis are not depleted.

Renewal takes time and does not always lead to a composition of the resource that coincides with human requirements. In such cases scarcity can develop, and there will be a call for active human intervention to speed up and direct the process. The way to avoid an existing or expected scarcity of a renewable resource is *production*.

3.2.2.   Some reasons for scarcity

The changes in society discussed in Chapter 1 (rapid population growth and a developing economy) lead to expanded and changed harvesting of benefits from the forest. This can in itself easily lead to situations of scarcity. The situation described in section 1.2 above indicates that for the next half-century there may be an unparalleled growth in the harvesting of forest benefits, leading to scarcities that can be avoided only by early action as regards proper management. The necessary managerial measures are, however, often hampered by factors such as common access, a lack of information, and financial problems.

Common access

In the early stages of social and economic development in a community, natural resources are often regarded as common property that everybody may use, but nobody feels responsible for them. In some parts of the world this is the case for quite extensive forest areas. It is often combined with a subsistence economy and a mainly rural population.

When the pressure on land and the consumption of wood in the area increases owing to expanding population, this situation leads to the destruction of forest resources and sometimes also to the depletion of productive land. To prevent a local scarcity of wood in such situations, it is necessary to create a linkage between the benefits from the forest resource and responsibility for it. This points to the necessity of local participation in forestry.

The same principle about the linkage of benefits and responsibility is valid for other benefits for other people. As Westoby (1987) says: "Unless we can win public authorities to the idea that the social services performed by the forest must be paid for by public funds, we shall progressively fall behind the level of investment required to ensure those services."

Lack of information

Even in societies where social and economic development is more advanced and where fairly good control over forestry activities has been established, a scarcity of wood or non-wood products can occur. One reason for this might be lack of information. This can of course mean a lack of knowledge about the forest resource and about management possibilities. However, the problem is far more complex when estimates of future or distant changes in human needs have to be made.

Financial problems

Developing the forest to a state that increases the output of scarce benefits will in most cases entail investments, and very often these investments take a long time to show a profit. In less developed countries there is often a very pressing need for investments linked to immediate problems. It is difficult to find the financial resources for long-term investment, for example in the production of wood (particularly industrial wood for future markets).

Looking at all tropical forest, the area of successful plantations up to 1990 was rather modest – about 1.5 percent of the forest area (mainly established on abandoned agricultural land). In spite of this, Palmberg-Lerche (1993) describes the situation as follows:

> The overall increase in the total area of plantations in the tropics between 1980 and 1990, was 150%; . . . there was a considerable, relative increase in non-industrial plantations in the form of woodlots, line plantings and agro-forestry: while the area under industrial plantations doubled in the tropics over the past decade, the area of non-industrial plantations almost tripled . . . . The overall share of industrial plantations in the tropics was at the end of 1990 about 35%.

The determination of whether this rate of expansion of industrial plantations will be sufficient to avoid a scarcity of industrial wood in the tropics will need further analysis.

Sedjo (1983) points to representative examples that indicate that the internal rate of return (IRR) of plantations in the tropics on average is much higher than that of plantations in the temperate zone – 14–24 percent in the tropics as against about 5 percent in the Nordic countries. ("Plantation" in this terminology does not necessarily mean an area where all trees are of the same species and planted in straight lines. It means that trees are planted with the aim of improving the harvesting of benefits from the area.) Discussing tropical America, Palmberg-Lerche (1993) says:

> Although disappointing as far as expectations are concerned, yields in forest plantations greatly exceeded those reached in natural forests. . . . This, coupled with other positive factors such as ease of access and uniformity of the raw materials produced in forest plantations, has led to the fact that e.g. in Brazil, where plantations represent only some 2% of the forest area, 60% of total industrial wood production is plantation based.

In many places in the tropics a well-planned increased planting programme could reduce future logging in natural forest. This would be more likely to happen if all the positive values of such a programme are taken into account when investment decisions are made.

In the preparatory work for the UN Conference on Environment and Development (UNCED, 1992a) it is stated that total forest-related official development assistance (ODA) in 1990 was USD 1354 million, and the paper continues: "A level of about four and a half times the present level of ODA will therefore be necessary to bridge the gap."

3.2.3.   Is there really a scarcity of forest products?

The scarcity discussed here has to do with products for consumption. They come from renewable resources, which could be managed to produce increased quantities.

Non-wood products

Many non-wood products do not appear in statistics. Particularly in less developed societies, a large number of local plants and fruits, wildlife, and fish are used for food, fodder, medicine, and other necessities for daily life. As the population grows and other forest products (timber and pulpwood) gain priority, the availability of some special plants and animals will be reduced. In most cases, substitutes for the disappearing products can be found in the market, but in a subsistence economy this does not always mean that they are available for the consumer. Local scarcity can then develop until economic development – which caused the problem – has lifted local purchasing power to a level that permits substitutes to take over.

Some non-wood products – together with local wood products (shingles, poles, etc.) – are important revenue earners with the potential for future growth (see Forest Department, Sarawak, 1990, and ITTO, 1990). Further development and marketing of these resources could help many societies overcome periods of scarcity, particularly in the first stages of economic development.

Editor Per Wegge (ITTO, 1990) concludes in his preface: "In spite of the unquestionable importance of non-timber products, little systematic research and development have yet been conducted to integrate these commodities in a more holistic approach to tropical forest management."

Fuelwood

Mainly on the basis of FAO statistics, WCED (1987a) states that "[a] major problem arises from the growing scarcity of fuelwood in developing countries. If trends continue, by the year 2000 around 2.4 billion people may be living in areas where wood is extremely scarce." WCED (1987b) estimates that the percentage of total energy supplied by fuelwood is: in Africa – 76 percent; in Asia – 42 percent; and in Latin America – 30 percent.

In discussing renewable resources of energy, WCED (1987b) states:

> One fact, however, seems certain. While we cannot go on using fuelwood and other biomass sources the way we do today, no significant substitution seems possible in the near and even medium-term future. Given this immediate and basic need for domestic fuel, and the low level of substitution possibilities, it seems, at least in the short term, that the only way out of this problem is to treat fuelwood like food, and grow it as a subsistence crop.

Collecting fuelwood, Peru. © sebra-film co.

This is a clear example of scarcity, and it seems to be rooted in all three factors discussed in section 3.2.2.

Industrial forest products

Paper is an extreme example of growing consumption (as described in section 1.2.3). For the time being, the recycling of used paper supplies part of the growth in demand. As long as recyled paper continues to contribute to production, the demand for wood as a raw material will experience relatively moderate growth. However, there are technical and economic limits to this increase. As these limits are approached, it seems likely that a rapidly increasing supply of wood will be needed.

Several analyses have been published on future trends in the production and consumption of wood – in particular industrial wood. Most of them are regional rather than global, although some global aspects are covered. This of course makes them of limited value in global assessments. One very important series has been *European Timber Trends and Prospects* (ETTS), which was first published in 1953 and whose fourth edition appeared in 1986 (ECE/FAO, 1986).

Another difficulty in drawing conclusions about sustainability from existing analyses is that they do not include the whole period of expected rapid population growth. For years such calculations have tended to have the year 2000 as their horizon, although ETTS IV (ECE/FAO, 1986) projected some regional aspects (consumption) up to 2025. Global analyses should probably attempt to look as far ahead as 2050 in order to encompass the expected growth in both population (see Fig. 1 above) and economy, and to account for the long time it takes for many important tree species to produce useful wood. Wood could be produced in a shorter time, and that will often be necessary, but it will in general need changes to natural ecosystems and the introduction of exotic species. Plantations on an industrial scale will also need time for securing suitable locations, and for planning, financing, organizing, etc.

The global situation was summed up by Peck during the 10th World Forestry Congress in Paris. His figures include wood as a raw material, industrial wood residues, waste paper, and fuelwood. In his conclusion he says:

> This paper presents a view of the global prospects for the consumption of wood and its products, as depicted by two authoritative sources . . . . The forecasts are unanimous in predicting further growth in the consumption of wood and its products up to the year 2000, and those that look further into the future do not expect it to stop at that point in time. The additional volumes that will be used are substantial – of the order of 60–80 million m³ EQ [cubic metres, equivalent volume of wood in the rough] a year. Only a modest part of the increase can be provided by the recovery and use of wood residues and waste paper; the bulk will have to be obtained from the forest resource.
>
> (Peck, 1991)

It should be noted that what is discussed here is the long-term trend and not short-term market changes.

The *UN–ECE/FAO 1990 Forest Resources Assessment* (ECE/FAO, 1993a,b) offers a better basis than was previously available to judge the risk of a future scarcity of industrial wood.

3.3.   Low income

3.3.1.   Income and sustainability

Most people will have a personal desire to increase their income, and people who make their living, wholly or partly, from the forestry sector of the economy are probably no different from others in this respect.

In many countries, forestry and forest industries are important cornerstones of the economy. Taking Sarawak, Malaysia, as an example: in 1993 as much as 40 percent of the government's revenue was estimated to come from this sector, the share of GDP being about 20 percent. For countries of this type, the level of income from forest resources is of paramount importance.

There is growing awareness that income growth and GDP per capita are not necessarily adequate expressions of human welfare (Worldwatch Institute, 1991). Although more complex indicators have been developed, income is nevertheless a basic part of these indicators, particularly for poor societies with few alternative resources.

The question frequently arises of how income growth can be combined with the desired sustainability. This has been taken up by the WCED (1987a):

> If large parts of the developing world are to avert economic, social, and environmental catastrophes, it is essential that global economic growth be revitalised. In practical terms, this means more rapid economic growth in both industrial and developing countries. . . . The Commission's overall assessment is that the international economy must speed up world growth while respecting the environmental constraints.

To speed up the use of resources and still achieve sustainable development does not at first sight appear possible. It is important, however, to distinguish between the various types of resources. As a basis for discussing sustainable increased income from the forests – as a renewable resource – it might be useful to look at some reasons for what we could call suboptimal income.

3.3.2.   Some reasons for suboptimal income

Underutilization of resource capacity

ECE/FAO (1993a) indicates that the volume of wood felled from exploitable forest in industrialized countries makes up 70–80 percent of the net annual increment. There may be sound technical or economic reasons for this. It might also be the result of good planning for sustainable development or of deliberate decisions based on priorities other than income generation. It is, however, clear that inadequate management often leads to non-optimal utilization of the forest, resulting in suboptimal income – either immediately or in the future. Such problems are often more pronounced in less developed countries where income generation is a most pressing need.

A lack of funds

A lack of funds for profitable investment in silviculture, logging facilities, or industry is very often a limiting factor for income generation, particularly in developing countries (see section 3.2.2 above).

Unsatisfactory adaptation of production to changing markets

Changes in society, as described in Chapter 1, will necessarily be followed by changes in the market for forest products. As mentioned in section 3.2.3, many forest products need considerable time to develop from idea to finished product. Each product has its own requirements as regards tree species, dimensions, and qualities. When planning forest management it is therefore useful and sometimes necessary to have an idea about future markets.

Changes in the consumption pattern for raw materials in the forest industry during the 1970s in the relatively well-developed and stable European market are studied in ETTS IV, with the following conclusions:

> The main changes between 1969–71 and 1979–81, for Europe as a whole, may be summarised as follows:
> - Consumption of logs (+10.6%) grew more slowly than that of pulpwood (+20.7%);
> - Consumption of coniferous logs grew steadily (+14.4%) while that of non-coniferous logs fell in the Nordic countries and EEC(9), but rose in Central and Southern Europe;
> - Among pulpwood assortments, consumption of coniferous round pulpwood stagnated, with a fall in the Nordic countries, while that of non-coniferous round pulpwood and especially of residues and chips grew strongly;
> - Consumption of all paper making fibres grew significantly over the decade (+30.4%). Growth was much faster for waste paper (+58.7%) than for pulp (+20.3%).
>
> (ECE/FAO, 1986)

In some parts of Europe the relative change was often larger. In less developed countries that are succeeding in their struggle for economic and cultural growth, consumption patterns can change much more rapidly (see section 1.2.3). The main trends in future markets are important as a background for planning forestry and forest industries if income generation is an important part of the policy.

A lack of appropriate technology – or the transfer of unsuitable technology

The transfer of unsuitable technology from places with different natural or socio-economic conditions can easily reduce income considerably. This can happen both in logging operations and in industrial development.

In less developed countries, methods of logging too often are copied from countries where labour costs are higher and machine costs lower. They may therefore in many cases be a waste of money and thus reduce income (Solberg and Skaar, 1986).

Reduced future income can also often be the result of using heavy machinery for logging. In a system of selective cutting, which is used all over the tropics in one form or another, big machinery on the ground inevitably destroys young trees and even the soil in which they grow.

One reason for high investments in roads and machinery, even when labour costs are low, may be that the logs are too heavy to be handled manually or with smaller machinery. In such situations, expensive and sophisticated technology might be necessary to generate any income at all. The choice of "man–machine system" is broadly discussed in the literature (see, for example, Sundberg, 1981).

A lack of adequately trained and organized personnel

Qualified personnel in forestry are extremely important both in private enterprises and in national forest authorities (NFAs). Prats Llaurado and Speidel (1981) have described the situation for NFAs in some countries in Africa, Europe, and Latin America in the early part of the 1970s. The examples seem to indicate that the ratio between forest area and number of NFA staff at that time was some 10–15 times higher in less developed countries than in industrialized ones. Some of the problems connected with the limited number of trained foresters are touched upon in section 1.3.1

above. The situation has certainly improved since the 1970s but, of the many aspects of forestry that NFAs must consider, the income-generating function needs appropriate attention if sustainable development is to be achieved.

3.4.   Geographical differences

The distribution of forest resources and benefits is very different from the distribution of population and needs. Forest resources differ a great deal between the various regions of the world, and also between districts within regions or countries. It has been indicated that geographical differences can lead to an unsatisfactory distribution of inputs and benefits, and thus to suboptimal aggregate benefits from forests. I shall first look briefly at some important factors in geographical differences and then try to identify the resulting conflicts of interest.

3.4.1.   Some important factors

The natural potential of the forest

The three fundamental factors can be said to be climate, soil, and topography. This can readily be seen if we move from the Arctic to the tropics, from the rain forest to the desert, from lowlands to mountains, from plains to rugged hills. The potential for the production of wood or other forest products varies considerably with these geographical differences. To take an example from one region: about 69 percent of Finland's land area is covered with forest, whereas the figure for Norway is about 27 percent (FAO, 1985). This difference is mainly due to the different topography of the two countries.

Population density

The density of the population greatly influences the structure and magnitude of the forest resource. FAO (1993) has statistical examples showing that within each climatic zone the proportion of land covered by forest decreases with increasing population density. Because the number of plant species is clearly determined by the size of the forest area, the figures suggest that a growing population leads to reduced biodiversity. At the same time, a growing population will normally need more of the benefits generated by the forest.

Research and education

The forest situation in many parts of the world is a result of the above two geographical aspects. However, research and education are increasingly an additional factor.

Economic capacity

The economic capacity of a society, as expressed for example by GDP per capita, provides information about the need for certain types of forest benefits – in particular forest products (see Fig. 2).

The availability of alternative resources

The availability of alternative resources for local use or income generation will – in combination with others of the factors above – indicate to what extent a society *depends* on benefits reaped from the forest. This dependence can be for the continuation of traditional life, for economic development, for the conservation of certain values in nature, for recreation, etc.

3.4.2.   Conflicting interests

Differences of the type mentioned above will naturally lead to varying views on how forests should be managed and how the resources should be used. The many possibilities for disagreement can largely be classified into two types:

(1) conflicts between the economic interests of two regions or countries, and
(2) conflicts between regional economic interests and general environmental interests.

Economic conflicts

Most economic problems are part of normal competitive activity. Some of them, however, have a historical background that hampers harmonious development.

One example is the *location of processing industries*. The economically more developed countries of the world initially established industries to produce useful commodities from their own forests. As time passed and more and varied raw materials were needed, they started importing logs from less developed countries. This contributed to economic growth in the countries exporting the logs and a desire in these countries to develop their own industry so that employment and added value could stay in the country. The benefits to the importing countries would thus be reduced.

The process of increasing industrialization in wood-exporting countries can sometimes be harmful to others and therefore create conflicts. It is, however, now broadly understood that this is a necessary development. International cooperation to smooth the road to a better situation will often be required.

The location of industry within a country can also give rise to conflicting interests: "We have the timber. We have export facilities. We need employment . . ."

Another example is *regional overexploitation* of forests leading to environmental and economic disasters in other regions (see section 3.1.2 above). Temporary overexploitation might well be justified for a country in a difficult situation, but possible negative impacts on other countries are seldom factors in the calculation.

Economy versus environment

The conflict between regional economic interests and general environmental interests has for some decades been a growing battle-field. There are clearly valid arguments on both sides, but it seems as if the geographical distance between opponents hampers real-istic discussion. Typical here is the fear often found in industrial-ized countries in the temperate zone of disasters caused by logging in tropical forests. Biodiversity and atmospheric carbon dioxide are the subjects of frequent debate (see section 3.1 above). It is certainly not difficult to find examples of forest management that can, with good reason, be criticized, but too often important infor-mation is ignored or misunderstood.

According to the FAO (1991), on the basis of data for 1989, the less developed countries (which mainly means the tropics) have 60 percent of the forest area and account for about 55 percent of the global roundwood harvest. In less developed countries, 80 percent of the harvest is for fuelwood and charcoal, whereas in in-dustrialized countries more than 80 percent of the harvest is in-dustrial wood. Of the industrial wood from less developed coun-tries less than 10 percent is exported, the rest being used locally.

The harvesting systems in tropical forests are not well known in other parts of the world. It is true that a great deal of damage is being done, but only big trees are taken out for industrial use. Smaller trees are left for future production. Clearcutting is not a part of normal forestry, but is done for non-forestry purposes such as agriculture.

Forest management in a distant country cannot be judged without taking into account the specific biological, sociological, and economic situation. A judgement based on only economic or only environmental arguments will inevitably be flawed.

There are good reasons for assisting less developed countries to improve the environmental as well as the economic aspects of their forestry, but to do it through restrictions such as a boycott will normally lead to the fostering not of "sustainable development" but of "sustained poverty".

3.5.   Social differences

Social differences can include differences in income, educational level, availability of medical and social services, etc. The reduction of such differences to an acceptable level is generally seen as a political goal. Particularly in less developed countries, among population groups with relatively low standards according to the above criteria, we will often find people living in the forest – usually in remote areas. To some extent the problems encountered are of a geographical nature, but they also have a social aspect and are therefore discussed under this heading. Many governments are trying in various ways to improve the situation for groups with least access to social benefits. Nevertheless, forest resources are often used in a way that widens the social gap in spite of the best intentions of the decision-makers. This typically happens in two different stages of development:

(1) In remote areas of some less developed countries where *hunting and gathering* is still the basic means of livelihood, any other activity in the forest may be experienced as a disturbance. Even if it is an activity established to create employment and income for the local population, it may sometimes develop too quickly. Traditional activities will then suffer while the people are not yet mentally or otherwise ready for fundamental changes in their way of life.

It is often apparent that a change towards a monetary economy is slowly occurring: an increasing percentage of the particular group of people are settling down more or less permanently; an increasing number go to school; and contact with other groups in the society is more frequent than before. It will then be important that the development is adapted to the local conditions. If changes happen too fast for the absorptive capacity to cope with, the result can be a setback for a particularly vulnerable social group.

(2) Logging – in particular of the large trees that traditionally are harvested in tropical forests – requires the use of *sophisticated technology*. The operation will then have to employ personnel with special training. As the operation moves on, the trained personnel follow and very little employment is generated for local people. As mentioned above (section 3.3.2), this high-level technology may under certain conditions be economically inappropriate. Too much of the income may be used for machinery and too little for personnel. In other situations, expensive equipment is necessary because the traditionally marketable logs are so big that heavy machinery seems to be the only solution. In both cases existing social gaps will remain or be widened (Jonsson and Lindgren, 1990). In such situations it is important to be aware of possibilities in the market for changes in forest management and logging methods.

Social differences based on different ways of life can also be part of the background when forest management for recreation versus production is discussed.

3.6.   Temporal differences

It was pointed out in Chapter 1 that long-term thinking has always been fundamental to forestry and that this thinking – particularly during the past half-century – has been changing. The main reason for this change has been the rapid growth and development of humanity. This has presented long-term thinking with a changing world where the future will in all probability demand increased services – quantitatively and qualitatively – from the forests. This brings us into the general problem of sustainability.

3.6.1.   Sustainability – the choice of concept

Nilsson (1991) discusses three possible concepts of sustainability:

- sustainable management of forest resources;
- sustainable development of tropical forests; and
- forestry for sustainable development.

The sustainable management of forest resources

The old concept of forest management (see section 1.1 above, in particular what Hartig said in 1804) was mostly related to the production of wood. Even if we include forest resources other than wood, the concept is still obsolete because it does not acknowledge the requirements of a changing world.

The sustainable development of tropical forests

"Sustainable development" is a term that – according to Nilsson (1991) – "is a fashion expression which is much used in today's international discussions. We find it in the pre-project report which is the basis for our discussions today, we find it in the outlines to a new forest policy for the World Bank, we find it in ITTO's Action Plan . . . . Sustainable development of the forests must mean that the ongoing deforestation can be arrested and that we in the future will achieve at least a small improvement of forest cover and forest quality, environmentally and econom-ically." This is a demanding objective, but it is still not linked to the actual requirements of the changing world.

Forestry for sustainable development

A concept that encompasses the totality of human requirements is far more complicated. This is a difficulty we cannot avoid if humanity is our target and our aim is to increase general welfare, or – in terms of Figure 3 – if our aim is to increase the aggregate benefit of forests.

Forestry for sustainable development can be linked directly to the definition formulated by the World Commission on Environment and Development (1987a): "Sustainable development is development that meets the needs of the present without compromising the ability of future generations to meet their own needs." The concept can also be directly linked to the definition of forestry given at the beginning of this chapter.

The task of managing the forest resource, with all its problems, possibilities, and competing interests (now and in the future), in compliance with this concept would appear to be beyond human capacity. It is. But that is the task, and the difficulties do not disappear if we close our eyes to them. We will have to simplify as carefully as we can. We will never be able to achieve an "optimal aggregate benefit of forests" that everybody can agree upon. But we can hope to get closer to it, and that is what we must work on.

The concept of sustainability used in this presentation will therefore be *forestry for sustainable development*. The consequence of this choice is that the production of wood, the maintenance of biological diversity, the stabilization of the global climate, etc. are means rather than objectives (cf. Nilsson). The objective is human welfare.

3.6.2. Geographical variation in future requirements

Natural conditions, traditions, and economic standards in a country are important factors in how forest resources are used (see sections 1.2 and 3.2.3 above). Looking only at the use of wood, it is clear that the pattern of consumption changes from country to country and from one period to another. ETTS IV (ECE/FAO, 1986) has discussed the consumption pattern for Europe and some other countries. By expressing the ETTS figures in m^3 EQ, per capita consumption can be compared with similar data for the world as a whole (FAO, 1985); see Table 4. Between regions or single countries the differences are of course much larger. To illustrate this, similar data for two very different countries – the USA and Nigeria – are also shown in the table. Europe – as one of the more developed regions of the world – has a higher consumption per capita of industrial wood products than the world on average. Only its use of fuelwood appears to be lower. In the USA, even the consumption of fuelwood appears to be higher than the world average. (The figures in Table 4 should be seen as approximations.)

Table 4. Consumption of some major forest products, 1980 (per '000 capita, m^3 EQ)

	USA	Nigeria	Europe	World
Sawnwood	932	73	405	212
Wood-based panels	156	5	87	29
Paper & paperboard	1091	7	373	157
Fuelwood & charcoal	402	920	136	337
Total	2581	1005	1001	735

The differences in nature and traditions indicate that large parts of the world will follow different directions of development than Europe or the USA has done. Demand for wood-based industrial products will continue to grow, but this growth will vary from region to region.

One very likely future growth in consumption is the growth of paper consumption in developing countries in the tropics. Because most of the natural wood in the tropics is more suitable for products other than paper, those countries are likely to be faced with the choice of either producing what they need in plantations or importing paper (or suitable wood) from temperate countries (if they can produce enough).

The future consumption of industrial wood in the various regions of the world will depend on many factors. My illustration will divide the world into only two regions, according to their current status: more developed and less developed countries. Three main factors will be taken into account:

- the number of people in the region (see Fig. 1);
- economic strength (see Fig. 2);
- the availability of alternative raw materials for commodities that "meet the needs" of sustainable development (because non-renewable resources are becoming increasingly scarce, renewable resources will have to take over where possible).

For this illustration I shall focus on the year 2050, partly because it is to be hoped that the most drastic population growth will be over by then, and partly because many important tree species need 50–80 years to produce useful wood.

Reliable estimates of the consumption of industrial wood so far in the future can probably not be made. As a basis for discussion it might be useful to sketch some alternative scenarios and possible figures for consumption of industrial wood (of all kinds) in the year 2050. Population growth is in all the scenarios assumed to follow the projections in Figure 1.

- *Scenario 1*. Consumption per capita in each of the two regions is the same as it was in 1975. There has been no economic growth. Non-renewable resources are available as in 1975.
- *Scenario 2*. The more developed countries have had very slow growth (0.15 percent p.a. on average), which raises consumption per capita from 0.897 m^3 in 1975 to 1.000 m^3 in 2050. This is assumed to represent moderate economic growth and

the continued availability of non-renewable or alternative resources.

- The less developed countries are assumed to reach the 1975 level of world consumption per capita in 2050. This means a growth from 0.086 m^3 in 1975 to 0.323 m^3 in 2050 (an average 1.8 percent p.a.). The corresponding economic growth is assumed to bring these countries to the world's 1975 average level of GDP by 2050.

- *Scenario 3*. World average consumption per capita in 2050 is anticipated to be the same as the average in the more developed countries in 1975. This corresponds to a growth rate of 1.4 percent p.a. These assumptions imply an average growth rate in GDP/capita of the same magnitude as the growth rate for consumption/capita.

The total consumption of industrial wood in the year 2050 would be as shown in Table 5, in comparison with 1326 million m^3 in 1975.

Table 5. Consumption of industrial wood, 2050 (million m^3)

	Scenario		
	1	2	3
More developed countries	1345	1500	–
Less developed countries	670	2520	–
World	2015	4020	8300

Actual developments from 1975 up to the present day seem to be in the neighbourhood of scenario 2. An estimate by the FAO (1991) for the year 2010 is significantly higher.

New technology, better use of the wood harvested, and recycling will help to reduce increases in harvesting. On the other hand, the projected economic growth – particularly in less developed countries – is lower than might be expected or hoped for. However, a preliminary conclusion based on the figures above must be that *a significant increase in wood production in decades to come – above all in less developed countries – will be a prerequisite for sustainable human development.*

In this connection it should be emphasized that in this period it is likely that there will be attempts to find substitutes for dwindling non-renewable resources. A closer analysis of these problems is an important task for international organizations and institutes with the expertise and other resources.

The problems indicated above cannot be postponed or ignored if a policy for sustainable development is to be formed. It would be an extremely "undesired managerial consequence" if development in poor countries were to be hampered or stopped by a lack of wood-based commodities – or because these countries had to buy the commodities at great cost instead of producing them themselves. Local and national action within an international framework is very much needed.

3.7.   Political choice and professional skill

Assuming that sustainable development is what we are trying to achieve and that the concept of forestry defined in section 3.6.1 is acknowledged, the ranking of benefits from the forest is a *political* process. It is a difficult process that needs the best possible information about human requirements and management options. If political intentions are expressed for one aspect, say biological diversity, it is important that the consequences for other aspects, say the problem of scarcity, are reasonably well clarified. This process has local, national, and global components whose priorities do not always coincide.

Taking as an example the global problem of sustainable wood production, one important question that has to be asked – and answered politically – is "What is our goal for wood production in the world during the next century?" Is it:

(1) to stop deforestation;
(2) to produce in the future about the same amount of wood annually as now;
(3) to produce about the same amount per capita as now;
(4) to allow room for some economic development – particularly in poor countries;
(5) to develop a renewable resource that can take over some of the tasks now performed by non-renewable resources?

It is a *professional* task – in cooperation with specialists in various fields – to clarify the possibilities, limitations, and consequences of management options. In this complex system of ecology, technology, economy, and sociology there are still many loose ends where an educated guess is the best information available. Research and resource assessments are vital for improvements.

Another important professional task is to develop managerial options where the "undesired consequences" are diminished or eliminated. This is a broad field of work where open and creative minds are important and where cooperation across professional boundaries is often a necessity. The remaining chapters will attempt to find possible ways of diminishing some of the undesired managerial consequences.

In the process of establishing better forestry for sustainable development, politicians and professionals have different roles to play. There is therefore an urgent need for mutual understanding and cooperation.

Chapter 4

Multiple use and priorities

Chapter 3 discussed a list of undesired managerial consequences (see also Fig. 3). The corresponding list of benefits offered by the forest varies from place to place, from period to period, and from one group of people to another. To establish a general and complete list of these goods and services would be an enormous task, if it were possible at all. For our purposes it is enough to provide an example that includes both conservation and production:

Protection: soil, water, atmosphere, biodiversity, landscape.

Production: timber, wood-fibre, fuel, food, fodder, tools, medicine.

Income: income generation, income distribution (geographically, socially, temporally).

Recreation

If more than one of these benefits is produced in a satisfactory way within a given forest area, there is already a type of multiple use. The complexity of human interests will often make it desirable to combine more of the benefits in the same area. In the following I shall look into some aspects of this process.

– Recreational use of the forest, Norway. © sebra-film co.

4.1.   Interdependence and compatibility

4.1.1.   Overlapping sectors of society

Forestry as defined at the beginning of Chapter 3 has more or less clear boundaries with other important sectors of society. How these borders are drawn is relatively arbitrary, but depends a great deal on traditions and local conditions. Three other sectors will usually be the closest partners with which forestry must cooperate to meet the needs of the society (see Fig. 5):

- environment
- rural development
- industry

The relationship between forestry and these other sectors can clearly be seen from the above list of benefits expected from the forest.

Environment

Important aspects of the environment depend on what happens in forest management (see section 3.1.1 above). However, environmental benefits are not always compatible with the available methods for meeting important needs related to rural development or industry.

Rural development

Agriculture is most often the leading instrument in rural development and in the production of food for an increasing population. Land use is a basic key to success in these fields. This can lead to competition over areas that for other reasons are needed for forestry, because in large parts of the world agriculture and forestry are incompatible on the same area. Under certain conditions in the tropics there is, however, an interdependence between agriculture and forestry. This has led to farm forestry and in particular to agro-forestry.

Industry

Sectors of industry that use wood as a raw material for production are, of course, strongly interdependent with forestry. Without wood they cannot produce, and without industry forestry cannot meet the needs of people for goods, services, and income. When scarcity occurs – or is anticipated – forestry may be unable to meet these needs without sacrificing benefits related to the environment or agriculture.

Forestry can produce the expected benefits only if the forest administration has a balanced relationship with its cooperating partners. There will certainly be many practical models for organizing cooperation. It is, however, important that the forest administration is not drawn so close to any of its "big brothers" that the holistic view is disturbed.

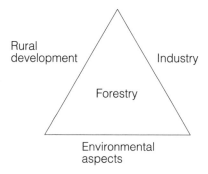

Figure 5. Forestry and its main cooperating sectors

4.1.2.   Conflicts between forest benefits

Some of the benefits mentioned above are interdependent. For example, all types of forest production depend on reasonable protection of the soil, and protection of the soil will need a certain cover of vegetation, in particular on steep terrain or in vulnerable tropical areas. Income generation will normally depend on some kind of production, even if protection also often needs activities that have to be paid for.

Other benefits are more opposed to each other. They have varying degrees of compatibility. If biodiversity is the main objective in a given forest area, the achievable yields from food production or traditional logging will be rather modest. Production for local consumption and income-generating employment will then in many cases have to be kept very low – or simply abandoned.

Multiple uses are possible and rational when management options consist of objectives that are either interdependent or have a high mutual compatibility. To be rational, multiple land use cannot mean that all management objectives will be fulfilled. However, the requirements of the multipurpose objective should, as far as possible, be continuously met by the overall management of the area. This means that:

- objectives with low mutual compatibility are best served if they are kept apart but given due attention in separate areas;
- production management systems that favour protection are given preference; and
- production objectives with high mutual compatibility (timber/ fuelwood) – or even interdependency – should be combined in the same area. Agro-forestry is one such system.

A general description of the degree of compatibility between various objectives in forest management cannot readily be given. Basic conditions differ a great deal and research in the field seems to be limited. It is, however, possible to establish a subjective picture that could be of some interest in the discussion about land-use planning and forest management in a given area. This is done by the ITTO Mission (1990) in its case study in Sarawak, and by Seip (1993) for certain conditions in Sarawak. With this as a background, Table 6 provides a rough subjective example relating to hill dipterocarp forest on moderate terrain and near longhouses (or villages) in Sarawak.

From the table it can for instance be seen that it was deemed possible to find an improved system for logging that would do less damage to soil and biodiversity and provide more local employment (this is discussed in Chapter 5). The table also indicates that, if protection is the primary objective, local employment will have to suffer, and under traditional logging systems only rattan production can provide a limited local income.

4.2.   Sustainability in a multipurpose forest

"Sustainable development" (see section 3.6.1) involves sustainability both in production and in environmental quality (protection) and is a natural basis for a discussion of sustainability in forestry.

4.2.1.   Production for sustainable development

In sections 3.2 and 3.6.2 it was shown that in a world in which the population will in all probability almost double during the coming 50–60 years, and where improved average income economy is not just a distant goal but in large parts of the world an ongoing process, increased production of commodities from renewable resources will be necessary.

Wood is one important such commodity. A relatively stable harvest of wood is important for employment and for a rational use of equipment in the forest and in wood-based industries.

| | Primary management objective | | | | | | | | |
| Secondary yields | Protection | | Local consumption | | | | Income generation | | |
	Soil	Bio-diversity	Food	Fuel	Medi-cine	Rat-tan	Trad. logging	Im-proved logging	Local employ-ment
Protection									
Soil	–	5	2	4	5	5	2	4	3
Biodiversity	5	–	0	4	5	4	2	3	3
Local consumption									
Food	3	1	–	4	1	0	0	2	4
Fuel	4	3	5	–	3	1	5	5	4
Medicine	5	5	1	4	–	3	2	2	2
Income generation									
Rattan	5	4	0	2	3	–	2	1	3
Traditional logging	1	2	0	4	1	1	–	3	3
Improved logging	4	2	3	4	1	1	3	–	5
Local employment	1	1	1	1	1	4	3	5	–

Arbitrary scale, 0–5: 0 = incompatible, 5 = perfectly compatible

Successive increases in production for the benefit of future generations seem to be inevitable to secure sustainable development. The forests of the world have many purposes, and the tricky task is to increase production and at the same time meet other objectives – often with a low compatibility with wood production.

The time factor is of importance as regards not only the necessary quantity of wood but also its quality or fitness for various purposes. Figures given in section 3.6.2 indicate that the consumption patterns change as a country develops economically. For less developed countries especially, the changes can be expected to be remarkable. The consumption of paper will, for example, in a particular period grow faster than that of most other wood-based commodities.

Sustainable development can be achieved only through a changing pattern of increased consumption of renewable resources. These trends are very difficult to predict with acceptable precision, but forest management cannot ignore them.

Forest management comprises a broad spectrum of methods. At one extreme it can involve careful selective logging of overmature trees and long intervals between one harvest and the next. At the other extreme it can involve smaller or bigger clearcuts, with natural or artificial regeneration and with frequent thinnings. There are of course many intermediate possibilities. In addition, there are many ways of carrying out logging and transport. The various methods have different results as regards sustainable production, employment, income generation, and environmental impact. Whereas the traditional selective cutting of big trees in tropical rain forest yields a yearly sustainable harvest that seldom exceeds 1 m³/ha, plantations normally yield 15–25 m³. In most cases accessibility and uniformity make plantations more economical to operate than natural tropical forest. In Brazil, where

Table 6. Example of assumed compatibility of certain forest management objectives near rural settlements in Sarawak

plantations represent only some 2 percent of the forest area, 60 percent of total industrial wood production is plantation based (Palmberg-Lerche, 1993).

Food is a commodity that in many areas competes with wood. In large areas of the developing world, shifting agriculture (slash and burn) is a dominant method of food production. This is a natural solution when land and local labour are the only available resources. In these countries population growth is often very high, so increased food production is a necessity. For a while this can be achieved by shortening the fallow period (the number of years between cultivations in the same area). If this process is carried too far (and it very often is), the result will be reduced fertility of the soil. This method of agriculture will then not be sustainable. The only solution for the low-income farmer is to burn more forest, and thus reduce the possibilities for reaping other benefits from the multipurpose forest.

Technical and financial assistance is required to establish sustainable use of natural resources. Trade restrictions may in many cases only increase the difficulties and block the road to sustainability.

4.2.2.   Sustainability in environmental quality

The basic environmental elements in forests everywhere are soil, water, vegetation, and fauna. Whereas water, vegetation, and fauna generally must be regarded as *renewable* resources (see section 3.2.1), soil could be called *re-usable*. It can be used again and again if properly managed, i.e. if erosion is kept to a minimum, and if its fertility is sustained or improved. Species or whole families within flora or fauna that are rare, vulnerable, or even endangered constitute a special case. In serious cases they must be regarded as *non-renewable* resources and handled with greater care than other species.

Traditional methods both in forestry and in certain types of agriculture will in many cases lead to erosion that on steep terrain can be unacceptable. Too short a fallow period in shifting agriculture destroys fertility. The quality and regularity of the supply of water for household needs, irrigation, fishing, and river traffic are also dependent on the minimization of erosion.

Endangered species will normally be threatened with extinction even without human influence. Careless forest management or other human influences, such as pollution (even from distant industry), can, however, increase the danger. Even endangered species must be subjected to a process of allotting priorities in which likely usefulness is also evaluated.

Not all harvesting of forest products is prejudicial to biodiversity. The indications are that moderate utilization in some cases gives forest vegetation greater diversity than occurs in undisturbed ecosystems. The species composition will have changed, however, and some rare or specialized species may be lost. Even narrow strip-cutting methods as developed in Peru seem to be compatible with biodiversity conservation (Blockhus et al., 1992). (See also section 3.1.1 above.)

Forest management systems that are geared more to high production can again have a different impact on vegetation and fauna. Clear-cutting followed by plantation will often result in one or a few tree species on a particular plot. If the plantation is dense, the vegetation on the ground can vary from very diverse right after planting to no visible vegetation at all in the dense stand as it approaches maturity. This is a periodic change. During the course of the rotation a great number of species – plants and animals – will be found and others will be lacking.

In addition to the conservation measures that might be implemented in production forest, some areas will need *total protection* – national parks, wildlife sanctuaries, etc. The production of the wood needed for global sustainable development could possibly be achieved by relatively moderate use of the production forest, but then a very small area would be left for total protection. It could also be achieved by intensive use of the production forest, leaving a larger area for total protection. It would seem that the combination of low production and large areas for total protection would not support sustainable development (see section 3.6.2).

4.2.3.   Long-term planning

It will appear from preceding chapters that forestry – as a means of sustainable development – is a long-term enterprise. The most obvious reason for this is the long time it takes for many valuable tree species to produce anything useful. Some species (Eucalyptus, Gmelina) can under certain conditions produce good timber in 5–10 years. Other species, both in the tropics and in the temperate zone, require a 50–100-year rotation. If the only objective of forestry were to produce wood – irrespective of the type of wood – we could theoretically make do with a much shorter rotation, and consequently a more limited planning horizon. This would particularly be the case in the tropics. But the result would be that the forest would increasingly consist of only a few quick-maturing species. This is not desirable from the point of view of the usefulness of a variety of forest products. And it is certainly not desirable for the environment.

In order to reduce future undesirable situations of this kind as far as possible, it is important to have a rough picture of future needs and what action is necessary now to meet those future needs. Long-term planning in this context entails planning not what *future* generations should do but what *we* should do now to give future generations the possibility of meeting their needs. The more choice we give them the better. But it is their choice. Their problem is that they are not here now to make the choice.

This type of planning does not just relate to the choice of tree species. Other aspects may be of even greater importance, such as land use, investment, and the level of harvest.

Land use

When areas are taken away from the forest, the benefits expected from the forest will be reduced, and this must if possible be compensated for in other forest areas. The problem can be dealt with much more satisfactorily if it is part of a long-term plan. An area that is set aside permanently for forestry purposes (whether production or protection) can be given the necessary attention, care, and investments for its purpose. If the future is uncertain, important investments are not likely to be made.

Investments

Investment will in general be needed in silviculture (e.g. plantations), logging facilities, industry, and institutions (research, education, administration). Such investments need coordination as regards timing and magnitude. As a basis for a coherent policy, long-term scenarios are increasingly used in several countries, for example in northern Europe.

One instance is a parliamentary paper produced by the Norwegian Ministry of Agriculture (Landbruksdepartementet, 1981). The scientific background was a report by Nersten et al. (1981), who analysed the consequences of various scenarios of investment and harvesting in production forest in Norway. There were four scenarios for future harvest and four for investments. Let us take as an example a scenario in which the harvest increases quite strongly through most of the coming century, and in which investment scenarios for the plantation area and fertilization are also discussed.

The type and magnitude of these scenarios are of course decisive for the volume of production and its increment at the end of the 100-year period. It appears that even if all investment stopped immediately, both the standing volume and the increment would go on increasing for a couple of decades because of investments already made. From then on they would decrease.

If investments in all production forest areas were raised to the level in what are regarded as well-managed forests, the volume and increment after 100 years were likely to be 40–50 percent higher than at the time of calculation. A still higher investment scenario was calculated to give about a 70 percent increase in volume and increment. All scenarios of course also give values for intermediate points in time.

Figures such as these and related figures have been used in policy formulation and budgeting. They can be useful in various connections, such as discussions about rural development, employment, exports/imports, etc.

Level of harvest

In the example above, the level of harvest as a function of time was decided *a priori*. In most cases scenarios of harvesting programmes are fundamental to the planning of sustainable development. Based on existing resources and appropriate investment programmes, the task will then often be to adapt harvest scenarios to future needs.

The research base for calculations of the type mentioned here is often weak – or even lacking. This is particularly the case where the species composition is diverse and where an objective is to sustain this variation. The situation in tropical forests will generally be very different from that in forests in the temperate zone. An educated guess is sometimes the only available estimate, but even that is better than nothing.

4.3. Priorities

In the preceding sections of this chapter, the intention has been to show that the forest has a multitude of benefits to offer, not all of which can be satisfactorily combined in a limited area. This is not always a problem. In many cases only one or a very few types of benefit are sought, so a management objective can easily be defined. Current developments, however, are clearly moving towards more complex objectives. This is particularly the case in regions of the world where there is great variety in natural resources and in socio-economic situations. To encompass all major management objectives, the term "sustainable multiple use" must then refer to a large area, such as a country or a geographical region. This situation, which is most likely to be found in tropical forests, is the focus for this presentation, without other areas being excluded from discussion of similar problems.

When two or more management objectives with limited compatibility have to be taken into consideration, it is necessary for some interests to be given priority in a given area. In the process of planning sustainable development, a series of criteria for assigning priority can be of value. These can represent economic, ecological, or social values. Because human welfare is the essence of sustainable development, different societies with different resources can have different interests and may therefore need different criteria for assigning priority.

Social groups can be formed in many ways. We could think of producers versus consumers, owners versus workers, or rural versus urban people. In the areas that are in focus here, a useful grouping could be:

- the local rural population, who are dependent on available land for their daily life;
- the population of the country or region in general, whose interests are presumably protected by government policy; and
- the population of the world at large, whose interests are expressed by a multitude of international governmental and non-governmental organizations.

It was seen in Chapter 3 that population groups of this kind may often have conflicting interests. It was also seen (section 3.6) that forestry for sustainable development cannot be built on a simple formula. Various diverse scenarios may satisfy the term "forestry for sustainable development", but different people may prefer different scenarios.

When political priorities have to be set, it seems justified to pay particular attention to local needs in the vicinity of established settlements – rural or urban. These local needs will have to be identified in each case. Rural societies in an early stage of development will often give a high priority to traditional products for consumption (for example, a large number of edible plants, medicine, etc.). Employment and increased income will, however, also be a particularly high priority for a majority of them. In urban societies there will also be consumer interests, often with a different spectrum of commodities (furniture, paper) from that in rural communities of the type mentioned above. As soon as trade and industry develop based on products harvested in the forest, even many urban societies will give priority to employment and economic benefits.

In all sorts of local societies some priority will be given to recreational benefits from the forest. It is not easy to make a generalization about what type of forest will serve this purpose best. Aspirations seem to vary a lot from country to country, ranging from wilderness to formal gardens. The most common desire may be that nothing should be changed – what I saw in my childhood, I would like to see again as an adult. In a living forest, there are of course limitations on the extent to which this can be arranged.

Further away from settlements, general national interests will naturally be dominant. These will be manifold, but will usually have a strong economic component. For some countries the forest is an indispensable part of their income potential, and it is vital for sustainable development to use it appropriately (see section 3.3.1).

The weights of priority suggested above are indicated in Figure 6. The figure illustrates a situation in which about half the forest area in a country is so close to established settlements (towns, villages, single farms) that local interests will need to be taken into account in forest management. In the part of the forest closest to a settlement (about 20 percent in this example), local priorities may dominate, but with increasing distance they will have to yield space to national priorities. This nearest part of the forest will generally – owing to the complexity of interests – need to be particularly in focus in policy formulation. Varying ownership and traditional rights present challenging problems for governmental policy.

Distance from an established settlement (% of forest area)

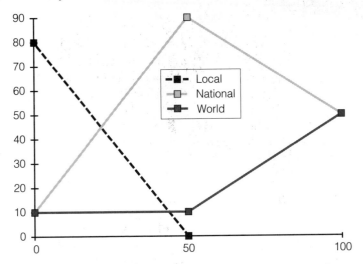

Figure 6. Local, national, and international interest in a forest area

The world at large has interests over the whole spectrum of benefits. The forest as a renewable source of wood, as a carbon sink, as a means of halting flood and erosion or maintaining biodiversity, etc. is of broad interest to the world as a whole. It needs to be taken into consideration to a certain extent everywhere. It seems reasonable, however, that benefits that are of greater interest to the outside world than to the local society should be given particular attention in more remote areas.

As society changes and forests are no longer capable of meeting all the benefits that are demanded of them, conflicts concerning priorities are growing. An important task must then be to search for managerial means of reducing conflict so that the aggregate benefit can be increased. Chapter 5 looks into this problem.

Chapter 5

Improving the aggregate benefit in a multipurpose forest

It has already been indicated (section 4.3) that various groups of people have their own policies and priorities as regards forest management. The components of these policies are discussed in Chapters 3 and 4. In this chapter I shall discuss some possible ways of implementing management in which the differences between group interests are diminished to a level where conflicts are reduced and sustainable development is promoted. This can be done partly by influencing the choice of priorities and partly by changing resource management in such a way as to reduce the gap between priorities. I shall summarize some central ideas and challenges connected with the complexity of forest use. I shall demonstrate possibilities as well as impossibilities, although both should be regarded as part of the real-life situation.

5.1. Policy and ways of influencing priorities

For a policy to be formed that takes into account conflicting priorities in a balanced manner, two things seem to be necessary:

(1) an expression of the priorities of each group, preferably with an analysis of the consequences; and
(2) ways of influencing these priorities and the final decision.

In the following I look into the problems of point (2) above.

 Take care! © sebra-film co.

5.1.1.   Instruments for balancing priorities

If we stick to the grouping used in Figure 6, there is now an *international* development (see section 1.3.1) towards an organizational structure with improved ability to protect the world's interests in points (1) and (2) above.

At the *national* level, policy has developed partly based on local experience and partly as a result of assistance and inspiration from outside – or worldwide – organizations (the FAO, ITTO, etc.). One difficulty in this development has been the increasing number of forest benefits that require administrative attention. When these benefits are placed under different administrative umbrellas – for example, splitting economic from environmental aspects – it is difficult to establish a holistic policy (see section 1.3.1).

Local interests and the way in which they can influence forest management are more complex. The necessity of improving local participation in all aspects of forestry will be discussed in sections 5.4 and 5.5. The reason for mentioning it here is to draw attention to the problem of local influence on national and international policy. People living in or near the forest will depend on forest benefits in various ways. The extent to which they can reap those benefits will largely depend on policy established in national or international forums.

If priorities are to have weight (for example, as indicated in Fig. 6), each group needs instruments to influence the decision-makers to be found at the various levels.

Such instruments might be:

- Information – research, dissemination of results, training, advice, extension services.
- Cooperation – agreements, organization, participation.
- Financial incentives – investments, grants, loans, development assistance, subsidies.
- Market development – trade liberalization, quality control,

analysis of consumption and production – with future trends.
- Laws and regulations – rights and limitations in the use of forests.
- Trade restrictions – boycotts, labelling.

The various interest groups have different sets of instruments available to them. Table 7 lists available and useful instruments that can be used by the various interest groups (governmental or non-governmental) to influence the choice of priorities and decisions in other groups. The list demonstrates the multitude and the complexity of the instruments, and reminds us of the various conflicts that might occur. However, the list also gives an overview of constructive possibilities and offers examples of actual tools in a challenging management situation.

Table 7. Instruments available to interest groups for influencing decisions

Level of decison-maker	Interest group		
	World	Nation	Local society
World	Information Cooperation	Information Cooperation	Information
Nation	Information Cooperation Financial incentives Market development	Information Cooperation Financial incentives Market development	Information Cooperation
Local society	Financial incentives	Information Cooperation Financial incentives Market development Laws and regulations Cooperation	Cooperation

Trade restrictions such as boycotts and labelling, which have been much discussed of late as instruments for achieving environmental benefits, are not included in the above suggestions. They are instruments with so many negative side-effects, which often are unknown to the initiator, that they should generally be avoided (see also UNCED, 1992b, para. 14). They tend to create more problems and conflicts than they solve, and therefore are likely to lead to a decreasing aggregate benefit of the forest. Responsible use of such instruments requires broad knowledge of the problems, possibilities, and consequences encountered in forestry in particular circumstances. "Certification", a newer form of trade restriction, is more positive than labelling, but it is still an instrument that could hamper the economic development that is indispensable in order to generate improved forest management. The corresponding compilation and dissemination of relevant information is of course of value, as long as it is not used in a discriminatory fashion.

5.1.2.   Financing incentives

To produce an increasing number of benefits in greater quantities the forest needs financial inputs. Investment for the owner's benefit will in general be procured by the owner. The owner's choice of benefits will vary from one owner to another, and may be different for private, common, and state ownership. Economic status – the ability to pay – is another important factor.

For benefits that have a lower priority for the owner but are regarded as important by other groups, one or more of the instruments listed in Table 7 may be required to influence the owner's priorities. Such benefits can be found within both production and conservation. Whatever instrument is used, a financial input is needed. Financial incentives will often be required to implement a policy that differs from what the owner/decision-maker would do if left to their own devices (ITTO, 1991). This happens in many situations – particularly where private ownership is dominant.

The financial base for these incentives is most often general taxation.

In other cases discussion may arise about who should pay – the beneficiary or the violator. Both principles may be useful, but definitions of benefits and violations are often difficult because they will depend on historical traditions and subjective emotions.

Let us take a few examples of what can be done if the aim is future increased wood production.

In many countries a part of the private owner's investment in the planting of forest trees is covered by public *subsidies*. The consequence is that the long-term profitability of the private investment will increase. More trees will then generally be planted until the marginal profitability is again the same as it would have been before subsidies. These subsidies are not given to improve the owner's immediate financial situation. The aim is to secure resources for future generations and to generate an immediate increase in employment and in a sustainable level of wood production. The investment is thus a joint venture between the owner and the general taxpayer, who will both derive some benefits from the investment, together with future generations.

A system that seems to have good results is what could be called a *compulsory investment fund*. It is described here as it is now used in Norway after a modest start in 1932. A certain proportion (normally 10 percent) of the gross income from the sale of industrial wood is placed in a bank account held by the forest authorities but with the owner's name linked to it. As long as the money stays in the bank, the owner gets no interest on it. The interest goes instead to the forest authorities, which, in consultation with representatives of the forest owners, use it for the general promotion of sound forestry (information, planning, etc.). Whenever the owner wants his money to make an investment in his forest (e.g. plantation) that is approved by the forest authorities, he receives it as tax-free income. In many cases this exemption from tax is of such great value to the owner – and the possibility for profitable investment so large – that he will choose to pay in more than the obligatory 10 percent to the fund. Here again, the owner and society both reap benefits together with future generations, but the system seems to be more effective than mere subsidies.

To illustrate what can be achieved by these two types of incentives, let us again take an example from Norway. In Chapter 3 (see section 3.1.2) it was noted that the growing stock in Norwegian forests almost doubled in the period 1920–1990. At the beginning of this period investment in planting and other silvicultural activities was negligible. It became increasingly clear, however, that it was politically desirable to increase the volume, increment, and harvest of wood. Both a compulsory investment fund and some subsidies were therefore established. Looking at the situation in about 1990, it appears that total silvicultural investment per year accounts for about 10–12 percent of the gross value of the harvested industrial wood. About a quarter of this investment comes from subsidies and the remainder almost exclusively from the compulsory investment fund. Without these incentives, a certain level of silvicultural investment would obviously have been reached, but nothing like the actual level. The level may be illustrated by noting that the number of seedlings planted annually is about 10 times the number of hectares in production forest.

One reason for the desire for more wood in the world is the effect this would have on supposed global warming (see section 3.1.2). World society is promoting sound forestry through development assistance, but, according to UNCED (1992a), not nearly enough (see section 3.2.2). In unofficial discussions the idea has been mentioned of letting a small proportion (say 1 percent) of the value of global oil production be used for reforestation in deforested areas. This could be to follow the principle that the polluter should pay the cost of the pollution. Through adjustment of the oil price the cost would be distributed among all consumers of oil. Implementation of a system of this kind might be more understandable and acceptable than increased taxes or transfers of funds from other development programmes.

Formulation of a balanced policy and the establishment of practical and acceptable means of implementing the policy are the first steps towards improving the aggregate benefit of the forest. In the following sections of this chapter I shall discuss some policy issues where it might be possible to reduce the gaps between interests by means of changed management.

5.2. Choice of land-use system

As time passes it is becoming increasingly clear that land is a limiting factor for human development. Looking ahead along the lines indicated in Chapter 1, how can we use land in a way that "meets the needs of the present without compromising the ability of future generations to meet their own needs"? This is of course a question with complex aspects far beyond the subject of this presentation. Some aspects of particular importance for forestry are briefly taken up here.

5.2.1. The process of urbanization

Towns and cities have normally grown up where people clustered together for reasons of natural resources – a good port, a waterfall as the basis for industry, or access to other important resources. Most often the existence of good agricultural land was a predominant reason. As a natural result of changes in society, urbanization is continuing and occupies land with a high potential value for plant production. In modern planning much is being done to direct the expansion of urban areas to less productive land. However, this is often costly, and immediate economic arguments easily win against long-term resource considerations. Agriculture and forestry must here sometimes yield their best areas.

Urbanization, together with industrialization and other related developments, also needs areas for roads, airports, dams, transmission lines, etc. to "meet the needs" both of today and of times to come. Again, agriculture and forestry inevitably lose areas.

Agriculture as a food-producing activity will naturally normally win the competition with forestry over the best land, so that forestry often seems to be left with areas that nobody else can use. This will in many cases be sound planning, but may also lead to disasters (see section 3.1.2).

Urbanization has yet another effect on forestry. Urban populations need nearby areas for recreation. This will influence the use of the forest area in many ways and limit other uses.

To the extent that it is possible to direct urban development and related activities to less productive land, the aggregate benefit of forests will increase. It may be expensive, but it will support the sustainability of development.

5.2.2.   Compatible management systems

Multiple use, with its possibilities and limitations, was discussed in Chapter 4. It was concluded that objectives that have a low compatibility with other important benefits should be satisfied in separate areas. Let us briefly look at some examples.

The example of assumed compatibility in section 4.1.2 indicated that the protection of biodiversity as a primary objective falls into the category of objectives that need some separate areas. Biodiversity can be reasonably protected under most management options except food production, but full protection can be given only in virgin forest that can be kept free from human influence, and even then natural changes are occurring (section 3.2.1). This underlines the usefulness of parks and reserves, but it also indicates that in most areas strict control of biodiversity has to compete with other important management objectives, and often loses.

The protection of biodiversity now seems to concentrate more on ecological functions, habitats, and vegetation types than on single species. This makes the problem more manageable. It also indicates that how big a part of the forest area it is desirable to set aside in order to protect biodiversity depends on existing variation and the vulnerability of the species in question. Each geographical or climatic region will have its own problems.

Soil conservation is important everywhere and cannot be limited to reserves. Its compatibility with wood production depends on how the harvesting of wood is done, and on the topographical and other natural conditions of the area. In areas with vulnerable soil on steep slopes, damaging erosion often occurs when heavy machinery is used for terrestrial logging. Examples are frequently found with traditional logging – particularly in the tropics. It then appears necessary either to avoid logging under such conditions or to find less damaging methods of wood production. Some possibilities for changing methods of wood production are discussed in section 5.3.

In some cases, fully compatible types of production are kept apart mainly owing to a lack of organization. For example, in areas where the tree species are useful both as timber for sawmills and as pulpwood, it may be that one part of the forest is harvested only as timber, and another part only as pulpwood. This inevitably results in wastage. Problems with transport are often a part of the reason, but better planning and cooperation between industries can sometimes lead to better utilization of the resource. Under more intensive management each tree can provide logs for various purposes. Multiple use is here an intensive and generally rational method of resource management, and the different types of production should if possible not be separated.

The choice of managerial system is therefore a professionally demanding challenge that requires knowledge, organization, and technology. These factors are important separately; they are, however, even more important when they are connected. It is only successful connection that will guarantee a positive outcome for society as a whole and for the specific locality.

5.2.3.   Area characteristics and production systems

I have discussed above how land has to be removed from agriculture and wood production. The area that is left can be used in different ways for agriculture, production forestry, or – under certain tropical conditions – the combination called agro-forestry. The choice between these systems will depend on existing features such as population, alternative income sources, and marketing possibilities. But it also strongly depends on the characteristics of the area – in particular its topography and fertility – and on what grows naturally in the area.

In all parts of the world agriculture started when the population was still small in relation to natural resources. As population increased, more areas were needed to produce food. In many places agriculture was forced onto land that by today's economic and ecological standards is totally unsuitable. In other places, forest covers land whose natural characteristics would suit modern intensive agriculture, but where there is a lack of population and infrastructure.

Technically, methods are now available to increase production from a given area in agriculture, and much has been achieved in this field – particularly in the more developed parts of the world. We can also in many places see a movement of agriculture to more manageable areas, but this is a very slow process.

It is beyond the scope of this book to discuss the problems of agriculture. It is, however, important for forestry to be involved in the development of stable land-use systems, where it is sensible to grow trees requiring a long rotation.

In the temperate zone, agriculture and forestry involve different types of production that in general cannot be mixed. In the tropics the situation is different, and it may be rational to use various systems of agro-forestry (see section 5.2.4). None of them, however, will exclude the need for separate agriculture and forestry.

The problem of how to use the area for production may therefore be more complex in the tropics than it is in the temperate zone. Figure 7 shows suggested rational uses that indicate how the choice will be influenced by some characteristics of the land. (For more on land-use definitions, see the next section.) In addition, the boundary between forest plantations and natural forest depends a great deal on the trees that are actually growing there initially and what they can be used for.

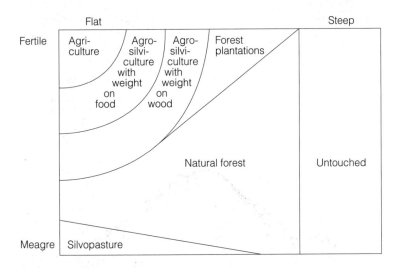

Figure 7. An indication of rational land-use choices in the tropics according to area characteristics

5.2.4.   Developing sustainable agriculture

The dominant form of agriculture in the tropics is still shifting agriculture (slash and burn). A few hundred years ago similar systems were also extensively used in parts of the temperate zone.

At a time when the population was low in relation to the available land, and other resources (e.g. fertilizers) were scarce, this was a rational use of the land. The system is that in a forest area the trees are felled and the area burnt. The ash serves as a fertilizer

Agro-forestry, Thailand. © sebra-film co.

for a couple of years, and in that time agricultural activity can be carried out in the area. Then a new forest area is cleared, burnt, and cultivated in the same way. When a cultivated area is abandoned, trees will again take over. Their roots will go deep and take up plant nutrients, which are brought to the surface and spread by the falling leaves. Over a period of 10–20 years the fertility of the ground is in this way rebuilt, and the area can again be cleared, burnt, and used for a couple of years for agriculture.

As the population grows, more agricultural land is needed each year. As long as the agricultural methods do not change, this requirement can be met only by either shortening the fallow period or bringing a larger forest area into operation. Both methods are used, resulting in less fertile agricultural land and a reduced area for wood production. The fallow period, which is when the trees grow and add fertility to the area, is now reduced to 3–5 years in many places, meaning that agricultural production will not be sustainable without improved methods. At the same time, this shifting agriculture has been forced to occupy areas that are inconvenient and vulnerable to erosion.

Two things are being done to improve the situation:

(1) the introduction of permanent agriculture in areas where this is possible (Fig. 7 indicates that this must be in the best parts of the area);
(2) the introduction of improved agro-forestry methods.

Agro-forestry can be defined as a group of land management systems that combine trees simultaneously or in sequence with agricultural crops and/or domestic animals on the same unit of land. The combination of trees and agricultural crops is *agro-silviculture*; the combination of trees and farm animals is *silvopasture*.

Traditional shifting agriculture can be regarded as a form of agro-forestry. It can be greatly improved by the deliberate use of trees that are efficient soil-improvers and produce useful and marketable products.

Farm animals in the tropics are mostly found in areas where humidity and temperature are not too high. That will often mean highlands where the forest is open, such as in natural savannahs. To improve the situation and achieve a more intensive state as silvopasture, the choice of tree species and the density of forest are important factors. In addition to the desirable properties mentioned above for tree species, trees in silvopasture should have branches or leaves that could serve as fodder. To make it possible for them to get through the juvenile stage without being eaten by the animals, a good deal of fencing will be necessary. This is costly, which can make the profitability of the improvement low and therefore less attractive. Living fences are sometimes used, and they can also produce fuelwood.

In more humid and warmer climates, where shifting agriculture as described above is the dominant rural activity, agro-silviculture can give good results both in economic terms and as a means of securing sustainable and efficient use of the land. Two main systems of agro-silviculture should be mentioned:

- *Alley-cropping.* Agricultural crops are cultivated under rows (alleys) of trees, so that trees and crops share the same area at the same time. This gives a fertilizing effect every year, but does not suit all crops (because of shading, root competition, etc.).
- *The Taungya system.* Here trees and crops follow each other on the same area. This method is relatively easy to introduce because it is a direct improvement of the most common form of shifting agriculture. Crops that are difficult to cultivate in alley-cropping will often fit better into a Taungya system.

Alley-cropping will in general be oriented towards food as the main product, whereas the Taungya system is more oriented towards wood production. Further information about agro-forestry can be found in publications by the International Council for Research on Agroforestry (for example, ICRAF, 1983).

5.2.5.   The productivity and sustainability of the land

Some 200–250 million ha of land in the tropics are now forest fallows – or secondary forest – after having been used for shifting agriculture. Some areas have been abandoned as unsuitable; others will be used again. By introducing suitable agro-forestry methods, production can be increased and land saved for forestry. Some of the secondary forest will provide useful products. Other areas will need to be planted with other species to be of any value. Left alone, these areas would take generations to restore biodiversity of the type they once had. It is preferable to establish plantations here instead of taking areas from the primary natural forest.

When farmers look for new land to cultivate, we can assume that they pick the best land available. The areas with forest fallows will therefore be among the more productive and accessible areas in the tropics. If for a moment we make the unrealistic assumption that we could replant the whole area with some suitable tree species, this alone might produce something in the order of 3–4 billion m^3 per year. That is considerably more than the world's annual consumption of industrial wood today. As the fallow area is now used, it has very low production of food and produces practically no industrial wood. This points to the importance of improving the use of such land to secure the sustainable development of humanity. There could be various kinds of use, and part of the area could be used for forest plantations.

The development of intensified and sustainable agriculture is not only necessary for food production but also of the greatest importance for the production of wood.

5.3.   Changing methods in wood production

It has been pointed out earlier (Chapters 1 and 3) that methods of wood production change from place to place and from time to time. This applies to the whole production chain: forest management, harvesting techniques, and industrial processing. Technical changes over time now seem to happen faster than before. This is due to research and general technological developments. It implies that the industry can rationally use an increasing proportion of the available wood, and that logging can be done by an increasing number of alternative methods. Rapid changes are more difficult in forest management, where each action is generally based on certain assumptions about the future – often a distant future – and cannot be changed if the assumptions are wrong.

To satisfy human needs and requirements, wood production will always play a dominant part in forestry. Increasing production will be necessary, and the challenge is to effect this with less damage to environmental and other values. This will involve the whole chain mentioned above:

- *Silviculture* will have to look into the future to estimate how much and what type of wood will best serve humanity in decades to come. Progress in the industry and in logging methods can generally be expected to give silviculture some freedom of choice, so that environmental values can be reasonably respected.
- *Harvesting techniques* must look for less damaging methods as regards both growing stock and the environment. Wood of the correct dimensions must be delivered to the industry at the right time and at a reasonable price.
- *Industrial processing* must look to existing and future human requirements and possibilities established by silviculture and harvesting. Its connection with the environment will often be

Fuelwood production (Ipil Ipil), The Philippines. © sebra-film co.

in the field of pollution, but it will also need to develop methods for using the raw material that fit in with an environmentally sound silviculture.

These three links in the production chain are interdependent. Together they can form an important instrument for promoting sustainable human development. To serve human development it is extremely important that the level of technology chosen fits reasonably well with the socio-economic situation (Jonsson and Lindgren, 1990; Solberg and Skaar, 1986; and Sundberg, 1981).

To illustrate the situation described above, let us take an example from tropical forestry where the problems and possibilities are relatively large. The example is picked from South-East Asia, where development at the moment is going very fast, where the natural forests have a relatively high percentage of industrially useful wood, and where the actual level of wood harvested in some countries may on occasion exceed the sustainable level.

Will it be possible to raise the sustainable level of wood production and also have a positive effect on the environment? Is it possible to take into consideration that future needs may be different from present needs as regards the choice of species?

Part of the answer lies in an increased planting programme on forest fallows (section 5.2.4), where annual yields in volume terms of industrial wood in plantations will be much higher per hectare than yields in natural tropical forest (section 3.2.2). A stronger growth in the consumption of products such as paper than in the use of timber is expected (section 1.2.3). If the policy is to secure at least a part of this future consumption from the country's own production, this can be achieved by plantations with adequate species.

More problematic is the question of what can be done in the natural forest. Traditionally the industry uses only relatively few species and trees of large dimensions. This goes back to when industrial forestry started in the tropics and wood was abundant.

Only the best of the best was harvested. Because of the big logs, heavy equipment is required for ground logging, and this frequently results in damage to other vegetation and the soil. The damage done by logging can be severe, but it rarely has anything to do with deforestation.

Two recent developments would appear to promise better solutions:

- helicopter logging of big valuable logs has been shown to be economically feasible and to do much less damage than traditional ground logging (Chua Kee Hui, 1993);
- there is growing interest in smaller logs in the market.

Sample plots in tropical forest show clear signs that a system of forest management where the trees are harvested when they reach a diameter at breast height (DBH) of 30–40 cm, instead of the present 60 cm or more, will sustainably produce more wood per hectare and year. The indications are that production of valuable species may be more than doubled (Seip, 1993). The smaller logs can be brought out with lighter equipment, doing less damage and generating more local employment.

The impact on biodiversity of such changes in forest management is unclear. The changes will have to be introduced gradually and at this stage are probably relevant only in locations requiring relatively short transport distance. Done carefully, it should be possible to avoid dramatic changes. Higher production will also make it possible to increase protection in other areas.

There is no reason to believe at the moment that primary natural forest in the tropics should be felled in order to establish forest plantations. This may come up as a conflict in the distant future, but it should not intrude upon the discussion now.

In the above example from South-East Asia it is clear that coordinated planning of forest management, logging, and industry is necessary to meet future needs. It is likely to involve considerable changes in all links of the production chain, but it could lead to

Pit-sawing, Honduras. © Hans Kristian Seip.

Logging of eucalyptus, Brazil. © sebra-film co.

significantly greater production and better protection of the environment.

5.4.   Broadening the spectrum of choices

I have discussed some ways in which forestry could be improved to support sustainable human development. Such improvement would be of value to humanity in general, but would be of particular importance for certain target groups. Rural people living in or near the forest will clearly be in focus. In a less developed country they will generally be among the poorest of the poor, often without any kind of cash income. From hunting and gathering, they have moved into subsistence farming with no other resources than a piece of land and their own ability to work. The subsistence farming that is normally practised is shifting agriculture, which provides full employment in certain periods, but at other times such people are looking for other income possibilities.

In some of these families one or two members can take full-time employment either in nearby forest (if they are trained for the work) or in a factory or other more urban establishment away from home. Other members of the family – men or women – will have to stay at home to do the farming. For these people a part-time job, where they could regulate their own time, could be the starting point for a development that could lift them above the poverty line. This would require a broadening of the spectrum of activities near their home. Could forestry provide part of this spectrum? An example is suggested in Figure 8.

The area that is currently used for shifting agriculture could be rearranged for other uses, depending on the characteristics of the area (Fig. 7). This would mean intensified management on much of the area. The steepest part might be so vulnerable to erosion that it should be left untouched, or at least very carefully used. Such places can sometimes be used to grow useful plants that do not need much light (rattan, edible plants traditionally harvested in the wild). Fuelwood could also be harvested.

In the forest outside this area, a zone could be set aside for industrial forestry operated by the village people. This would need management of the type described in section 5.3. Big logs could be taken out by helicopter and the smaller ones by lighter tractors – maybe a pedestrian controlled tractor – which could be operated by local people after some training. In some cases oxen could be used. Another possibility could be to let the villagers deal with the bigger logs by splitting them on site (using pit-saws or portable sawmills) and then bringing them out with light equipment.

In both areas silvicultural work to improve production would be needed. This could also be a source of income for the village.

The wood harvested in this way could either be transported to existing industry (sawmills, plywood factories, chip mills) or processed in the village. This could be done by building a small sawmill to produce sawnwood of a quality sufficient for the local market. Other "cottage" industries could also be established with wood or food as their basis (e.g. furniture, fruit juice).

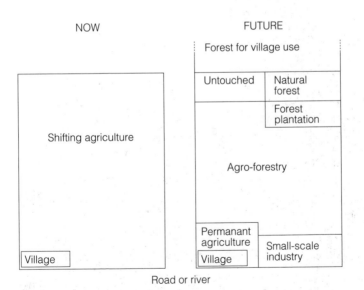

Figure 8. Example of a broadened spectrum of choices

Paper production, France. © sebra-film co.

The possibility of moving from "Now" to "Future" (Fig. 8) and the speed with which it could happen would depend upon the consistency of the policy and a willingness to organize the process. "Future" represents a goal. If this goal is widely accepted, we should work with that outlook and give priority to financial incentives towards that goal.

A development as described above would create a certain amount of employment and income in some locations. Further development in larger areas – such as a country – would need large-scale industry and a well-planned infrastructure. However, local development, for example as described above, will often be a necessary starting point.

The path of development – as from Now to Future in Figure 8 – can, and should, take many forms, because no place or society is a blueprint for another. The sooner we find a path that can serve *sustainable* human development, and start walking it, the better. In the rest of this chapter I shall take a brief look at some indispensable tools in this process.

5.5.   Organization

Governmental and non-governmental organizations (NGOs) at various levels were briefly mentioned in Chapter 1. NGOs play an important role as agents for group interests. The balancing of priorities in the search for improved aggregate benefit will, however, generally be the task of a governmental organization. This is the case both nationally and internationally.

5.5.1.   *National forest authorities (NFAs)*

The national governmental organization for forestry will generally have a threefold role as a planning, advisory, and executive body. These roles are fundamentally different but nevertheless closely connected.

- Planning includes the presentation of alternatives together with analysis of the limitations and consequences. This is important as a basis for political decisions. The NFA can also assist the private sector with planning.
- As an executive body the NFA implements political decisions. Within the framework given by governmental decisions, the NFA will necessarily have to make its own day-to-day decisions during the implementation process.
- The advisory role is a two-way activity. The national political level will need professional advice during the process of decision-making, and the local owner or user of the forest will depend on close contact with qualified advisers in their search for optimal solutions within the framework established by political decisions.

In performing its tasks the NFA must:

(1) cover all aspects of forestry. Its staff must be well qualified and able to take an overall view as well as being versed in most of the important elements. Outside specialists will often be brought in for particular tasks, but it is essential for the NFA to be able to put the elements together into viable management options.

(2) have a structure that enables the NFA to have close and direct contact with the political level as well as with the local users of the forest. Good and short lines of communication within the NFA, that is, between headquarters and field staff, are also important both to implement the general policy and to secure feedback for improving the policy.

A well-developed research and training system is necessary for any forest administration. In forestry – maybe more than in most other fields – local experience is essential. A certain permanence in staffing is therefore of great value.

5.5.2. People's participation

It has been mentioned above (section 5.1.1) that various interest groups need to be involved in policy formulation. In the actual planning and implementation of forest management the number of participants has to be more limited for practical reasons. It is, however, important that the local population in the actual area is involved in the activity. First, if the policy is to contribute to sustainable human development, the local population will be the most obvious target group whose interests should have high priority (see section 4.3). Second, a change in traditional management cannot be effectively implemented if people living in the area are strongly sceptical about it. They must feel secure that the new type of management will give them benefits that are at least as good as, and preferably better than, those provided by the traditional system. A governmental policy that includes increased benefits for other groups in society as well will have a reasonable chance of success only if the local people are partners in the process.

The local benefits will have various components, among which employment, participation in decision-making, and profits on sales will most often have high priority. These benefits can be reaped from the total process of forest management or from parts of it, such as:

- improvement of protection forest,
- production of non-wood commodities,
- wood harvesting and transportation,
- wood processing,
- marketing, and
- planning and administration.

For the local people to influence the situation in a beneficial way, some kind of organization is necessary. Internal cooperation between individuals in a group can be established at various levels of integration. Three examples will be mentioned here: coordinated activity, an association, and a cooperative.

Coordinated activity

Coordinated activity means that a number of individuals or families organize their activities (or some of them) in the same way. An illustration from Africa can be used here.

In a state-owned forest area the national forest authority built a village with houses of a better standard than was normal in the area. A shop, a school, and some medical services were also provided. The landless people who lived by shifting agriculture in the same area were offered these facilities and secure employment in the forest. Because the area had been used for shifting agriculture for a long time, forest plantations were needed, and the local people were able to do both the logging and the planting.

After logging in one sector of the area, people were allowed and even encouraged to cultivate agricultural crops in that sector. At the same time, trees were planted and the villagers who practised agriculture between the seedlings were responsible for replacing seedlings that died. As the logging moved from sector to sector, the agriculture and the tree-planting followed. The villagers' situation was clearly improved in comparison with what they had before, and the forest was changed from unmanaged to intensively managed with a mixture of permanent forest and agro-silviculture based on the Taungya model. It was a clear improvement of the aggregate benefit, and it seemed to succeed because the local people realized that they received significant benefits.

Associations

An association is where a number of individuals or families form an organization that enables them to act as a unit in certain situations related to their activity (or a part of it). The example here could be taken from several European countries with a large number of relatively small private forest holdings. Early in the 20th century the forest owners felt a need to join forces in their negotiations over wood prices with the industry. Generally this started locally and then grew into a national federation of forest owners. From the start the main task of such a federation was to deal with the marketing of wood for its members. Thereafter, the organization in some countries found it useful to expand into fields other than marketing, such as organizing a labour force to assist its members, extension services, and industries.

Cooperatives

Cooperatives in which a number of individuals or families join in a profit-sharing economic activity are much more common in agriculture than they are in forestry, possibly because the longer production cycle in forestry makes it more difficult to keep a continuous record of changes in values. But even in forestry we find various examples that would fit this definition of a cooperative. In countries with different traditions and political patterns we often find diverse types of cooperatives, and many of them seem to be successful.

One example can be taken from Honduras. After some years of what was called a "social forestry system", in which villagers living in the forest were given the chance to earn an income on the basis of resination, maintenance work, etc., several villages were involved in an integrated forestry development approach. Each village was given the opportunity to buy standing trees from the government within a certain area around the village. The trees were picked out and marked by the NFA, which also gave advice on how to use handtools and oxen for logging, and paid the people for silvicultural work. The villagers were also given assistance in negotiating prices with the sawmillers buying the logs. The project was reported on to the IX World Forestry Congress in Mexico in 1985, where it was stated: "This last stage has proven to have magnificent results since previous attempts in communities have led to an increase in average gross income by 400% after barely one year of work. Such results have clearly shown that rural development in Honduras can be based on forestry activities" (Barahona, 1985).

People's participation in one form or another can be established irrespective of ownership, although the practical solutions may often be different in state, common, and private forest. They will also depend on traditions, educational level, and available resources. Finding a model that fits the particular society can be demanding, but it is often the key, not only to local development, but also to improving the aggregate benefit of the forest.

Chapter 6

Policy and planning

There is a comprehensive literature covering problems of policy and planning in forestry, of which I shall specifically mention FAO (1987) and Nilsson (1992). Valuable checklists and planning systems can also be found in various guidelines for project formulation (for example, ITTO, 1992).

In addressing policy and planning here, I am attempting to establish a path through issues as they have been discussed in previous chapters. In doing so, I feel it is important to think of *sustainable development* as it is defined by the WCED (see section 3.6.1), as a *multitude of possibilities and interests* (see Chapter 3), and as necessitating *cooperation* (see Chapter 5). The process of planning is in itself an exercise in cooperation in which interest groups (individuals, organizations, countries) work together in search of the "optimal aggregate benefit". If confrontation takes the place of cooperation, no optimal solution will be found.

Policy and planning for a private forest estate or a concession area or a local community (Fig. 8) will be different from what is needed at the national level, where planning can be more concerned with trends and approximations, but is also often more complicated (see Fig. 9). In addition, the world at large needs to have a forest policy and certain plans. The process outlined in section 6.1 relates to the national level. In section 6.2, I attempt to pinpoint a few issues that need to be discussed as elements of a global forest policy.

← Necessity is the mother of invention. © Hans Kristian Seip

6.1.   The process of balanced national planning

To establish a balanced and effective national plan for forestry, at least three things are needed: the political will, qualified personnel with access to the necessary information, and a sensible way of organizing the work. The following suggested process is not the only possible method, but it may be a useful guideline to balancing the many interests. The process is outlined in Figure 10, and the main points are briefly discussed in the following.

6.1.1.   The initiative

A basic prerequisite for a successful planning process is the political will to improve the aggregate benefit of the forest (see section 3.7), and this must result in an initiative to start the process of planning. The political level can of course make use of available staff to formulate an initiative, but responsibility for its content must rest on political shoulders. The initiative should consist of the taskforce and its terms of reference.

Taskforce members and contacts

The complexity of the problem and the necessity of cooperation in the pursuit of a common goal – the optimal aggregate benefit – require broad representation and flexible minds. People with a central position in a forestry department (NFA) will generally form the core of the group. Depending on how well endowed the forest department is with experts in various fields, others can be called upon. Some degree of involvement will be required by people from other government departments, representing, for example, the Planning Commission, Land-Use Authority, Industry, the Environment, Research, etc. From the private sector, forest owners or concessionaires and environmental NGOs are possible participants. Most of these will fall into the group of contacts that will be called upon occasionally, but in some cases a limited number of such experts can be asked to join the taskforce in a more permanent way. It will also be important to maintain contact with the political level during the planning process. It may therefore be found practical to have one member of the taskforce from the political level. For a national plan, relevant data and policy in other countries or in the world at large will need attention. For less developed countries in particular, technical and/or financial assis-

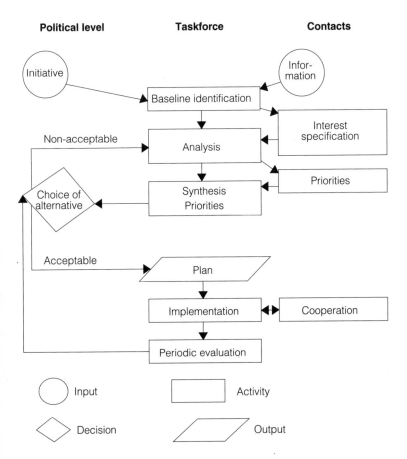

Figure 9. The process of balanced national planning

tance may be required. Organizations for bilateral or multilateral assistance can here be called upon to participate. The Tropical Forestry Action Programme (TFAP) is a natural contact in such cases for countries in the tropics.

Terms of reference

The initiative should also include terms of reference that point out some of the more important issues that, from a political point of view, need to be taken up. They may also establish limits on what problems should be taken up and on the time available for the process.

6.1.2.   Baseline identification

The work of the taskforce will necessarily begin with the collection of data and other information. The results will form the basis of the planning process, describing a situation that the plan will try to improve. This baseline will therefore also be used for comparison in future evaluation.

It should be emphasized that the need for this identification and its comprehensiveness have to be evaluated and defined in each case. Here I shall consider some possible items in general terms.

Forest resource assessment has always been the base for a forestry plan. It is, however, important that the assessment comprises information that is relevant and provides reasonable coverage of the various aspects of the plan. The accuracy of estimates will always be a question of cost and should not be driven further than the actual task requires.

Actual activities – both quantitatively and qualitatively – need a description. These are the items that may need to be changed in order to improve the aggregate benefit. Current forest management, such as level and methods of harvesting and level and types of investments, will generally be in focus, but so too will be industrial processing, protective activities, etc.

Current and expected benefits can be listed and described. This will need good cooperation with contacts within both production and protection. Social and political issues may be considered. The expected benefits will often be an improvement of what in Chapter 3 were called "managerial consequences".

General national policy must be taken into account as far as it is relevant for forestry. Land use, social equity, and economic growth are examples of political priorities that influence the national plan for forestry. The existence of alternative resources can indicate to what degree the national economy is dependent on income from the forest.

World forest resources and trends in the production and consumption of forest products are basic data for the discussion of future markets, possibilities for exports or imports, etc. It is an important task for international institutions to collect relevant statistics and estimate future trends as a background to national planning.

6.1.3.   Analysis

In the analysis a limited number of alternative programmes for the forestry sector are studied and described in a way that will enable the political level to make a choice. In this operation, close contact between the taskforce and the political level is desirable, so that uninteresting alternatives can be dropped at an early stage.

An analysis can be built up in various ways depending on which objectives are uppermost. In a national plan for a renewable resource like forests, it is natural to look into the sustainability problem first in order to find an approximate level of desirable activity. Building on information collected during the baseline identification (section 6.1.2) and on estimates of future changes in society (section 1.2), alternative long-term scenarios can be drawn up. Such scenarios are discussed in section 3.6 and Chapter 4.

Based on one or two long-term scenarios, the analysis can now consider "the roots of dissatisfaction", for example as they are shown in Figure 3 or discussed in Chapter 3. Examples of what can come out of this part of the analysis are given in sections 5.2, 5.3, and 5.4. A change in any of the "roots", or, conversely, the benefits, may influence other benefits. An improvement in one benefit should therefore be followed by an explanation of the consequences for other benefits.

To implement the basic scenario and each of the recommended improvements, some instruments will be needed (see section 5.1). Possible instruments and their costs should be discussed in the analysis.

6.1.4.   Synthesis

The results of the analysis can then be discussed with contacts outside the taskforce. When these contacts have had the opportunity to express their viewpoints and priorities, the taskforce will be able to condense the results of the analysis into one or two holistic scenarios with a corresponding plan for what has to be done in the coming 5 or 10 years. This synthesis, together with the comments and priorities of the taskforce and the NFA, then goes for political approval.

After discussions at the political level – in close contact with NFA – one of the alternatives may be chosen as the actual plan for the coming few years. If none of the alternatives finds approval, it might be necessary to do some further analyses in order to arrive at an accepted plan.

6.1.5.   Periodic evaluation

During the implementation of the plan, the NFA will often need to cooperate with external contacts to discuss unforeseen issues. Each year budgetary problems have to be solved, and after a few years a more general evaluation will be necessary to enable the political level to decide if the plan is still valid or if new analyses would be desirable.

6.2.   A global challenge

The need for a holistic global forest policy is increasingly recognized. The possibility of establishing an "Indicative World Plan" was discussed in the FAO in the late 1960s, but was abandoned. Discussions in various international forums have, however, contributed to the clarification of the many forestry issues of global common interest. A positive step forward was made in the "Agreements on Environment and Development" (UNCED, 1992b), in which a series of principles of great importance was drawn up. A global perspective on forest policy is also given by Sharma et al. (1992). Further developments in this field are bound to follow in years to come, and will be of great value to everyone dealing with national or international forestry problems. The UN Ad-hoc Intergovernmental Panel on Forests has an especially important task.

One particular element of global forest policy will be mentioned here because it is strongly related to some of the problems considered in earlier chapters.

Sustainability – as defined by the WCED (see section 3.6.1) – has been a basic principle in this presentation and is frequently included in the discussion. It is generally agreed that sustainability is a primary goal. When applying this principle in a national or regional plan, certain aspects must be kept in mind. The term "sustainability" is here related to human development, and therefore does not depict a static, unchangeable situation in the forest. On the contrary, changes in society (see section 1.2) will require changes in the structure and use of the forest.

One problem here is that if the forest is to serve human development (by providing near-optimal aggregate benefit), the structure of that forest must be planned a long time ahead. This planning for an unknown future is not easy. To enable nations to do their planning on the basis of the best available estimates of future demands on the forests, international institutions and organizations should intensify their studies of future trends. *The UN–ECE/FAO 1990 Forest Resources Assessment* offers a base for studying alternative programmes for the use and conservation of forests. Future human needs, however, must be estimated on a different basis, where the long-term trends are even more complex and uncertain than the growth and development of a forest. These trends are, however, important for forestry for sustainable development.

One such trend of importance for long-term global policy is the future availability of some finite or non-renewable resources. Wood – as a renewable resource – can be produced in large quantities if the necessary investments are made at the appropriate time. The physical and chemical properties of wood indicate that it may be used for purposes that are now covered primarily by non-renewable resources such as oil and iron. At the moment, wood is often too expensive in comparison with traditional resources, but when the non-renewable resources get scarce and therefore expensive, and wood is made available in greater quantities and is therefore cheaper, things may change. When will that be? And how should it influence the forest management in the near future?

There is an important discussion going on about how we could use less of the world's resources in order to safeguard sustainable development. It would be a much more important discussion if attention was focused more on how we can produce *more* resources in order to safeguard sustainable development at a more satisfying economic level. The role of wood in this connection represents a significant element. In this field, forestry could be decisive for human development and future life.

Let us now go back to "The problem tree" (Fig. 3) in Chapter 1. As we approach the 21st century, we find that the left side of the "tree" – the "inadequate generation of benefits" – has developed

into three goals that are all important but not fully compatible
(Fig. 10):

- environmental stability
- sustainable development
- profitability

The right side of the "tree" – the "unsatisfactory distribution of
inputs and benefits" – has developed not only into an unpleasant
situation but into an obstacle to the achievement of the best
possible combination of the three above goals. This can occur at a
national level, but let us here give it a global dimension.

Figure 10 indicates that if no outside influence is brought to
bear, forest management will have great difficulty combining the
three goals. In the figure the goals are indicated by rectangles,
illustrating that each goal can be reached by various methods of
management under various ecological and socio-economic conditions. A small square is common to all three goals. But, apart from
that, the environmental and profitability goals in particular seem
to require different management systems.

If we stick to the assumption that *sustainable development is the
primary goal*, the global forest strategy must aim at making as
much as possible of the two other goals compatible with the primary goal. Figure 10 illustrates how this might be done. As long
as the "unsatisfactory distribution of inputs and benefits" is not
improved, very little can be done to combine the three goals. Various instruments (see section 5.1) must be used to change the
situation. It seems appropriate here to apply the quotation from
Westoby (1987; see section 3.2.2) to the international situation:
"Unless we can win the public authorities to the idea that the
social services performed by the forest must be paid for by public
funds, we shall progressively fall behind the level of investment
required to ensure those services." Effective instruments are those
that correct the "unsatisfactory distribution of inputs and benefits".

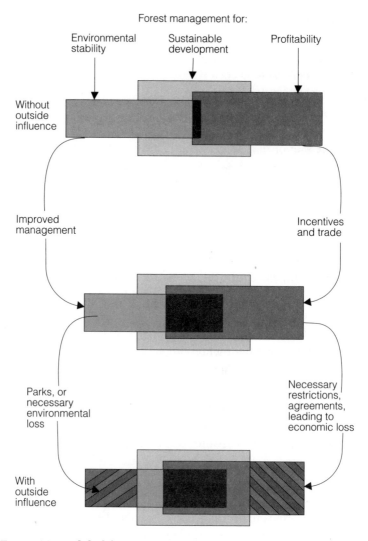

Figure 10. A global forest strategy

An example. A less developed country uses logging machines that cause environmental disruption (erosion, damage to trees and other vegetation). Better methods are known but are too expensive for the country's financial circumstances. An international organization that is aware of the situation and its negative impact on sustainability provides subsidies that bring the cost of the better method down to a level that enables it to compete with the traditional method. As a result, profitable forest management has to a larger degree been made to serve sustainable management, and the beneficiaries have shared the cost.

Similarly, if environmental interests have concluded that an area should remain untouched – and therefore be withdrawn from production for sustainable development – because the ecosystem is unique, improved management techniques may be developed that are satisfactory from an environmental point of view. A greater part of the vulnerable area is then brought in to serve sustainable development. Another solution may be to increase production in a nearby area by plantation, application of fertilizer, or otherwise. Again the input of information or funds often has to come from the outside.

The lower part of Figure 10 includes instruments aimed at eliminating forestry activities that are still incompatible with sustainable development. There will often be disagreement on what is tolerable, and it may be difficult to press somebody else's views on a sovereign government. This can generally only be done by mutual agreement and after a good climate for cooperation has been generated through the use of "positive" instruments as mentioned above. Again, cooperation is much more efficient than confrontation, and an improved distribution of inputs and benefits may be the best – if not the only – way to achieve a solution. There is a tendency to suggest unpopular instruments before more positive ones have been tried and a willingness to share the cost for common benefits has been demonstrated. This attitude makes it difficult to reach binding international agreements, or to establish an international organization with sufficient strength to advise and support the implementation of this process – both technically and financially. If parks and other areas that are being reserved for non-productive use can as far as possible be located outside areas that can be made profitable for wood production, this will serve sustainable development. A certain loss of values – environmental as well as economic – is, however, unavoidable in forest management aiming at sustainable development.

The optimal aggregate benefit of forests must be found within the framework for sustainable development. A strong global organizational pattern, based on cooperation instead of confrontation, and effective political instruments – even if they are expensive – are prerequisites for solutions near the optimal aggregate benefit of forests. It is an important national as well as international task to build that organization and to develop those instruments as soon as possible.

References

Barahona, J. E. (1985), "Social forestry system. The case of Honduras", Exposés, Abstracts, IX World Forestry Congress, Mexico.

Bennet, E. L., and Dahaban, Z. (1992), "Response of wildlife to different types of disturbance in Sarawak, and implications for forest management". Workshop paper. International Workshop on Ecology, Conservation and Management of Southeast Asian Rainforests, Kuching, Sarawak, Malaysia, October.

Blockhus, J. M., Dillenbeck, M., Sayer, J. A., and Wegge, P. (1992), *Conserving Biological Diversity in Managed Tropical Forests.* IUCN/ITTO. The IUCN Publication Service Unit. Cambridge, UK.

Bostrøm, K. (1986), "Deforestation, some aspects". Paper presented at Expert Group Meeting on Environmental and Socio-Economic Aspects of Tropical Deforestation (ESCAP), Bangkok.

Carson, R., (1962), *Silent Spring.* Fawcett World Library, New York.

Council of Environmental Quality and the Department of State (1980), *The Global 2000 Report to the President. Entering the Twenty-First Century.* Volume One. Washington, D.C.

Chua Kee Hui, D. (1993), *A Case Study on Helicopter Harvesting in the Hill Mixed Dipterocarp Forests of Sarawak.* Research report No. FE 2/93, Kuching, Malaysia.

ECE/FAO (1986), *European Timber Trends and Prospects to the Year 2000 and Beyond* [ETTS IV]. New York: United Nations.

ECE/FAO (1993a), *The UN–ECE/FAO 1990 Forest Resources Assessment. The Forest Resources of the Temperate Zones. Vol. I: General Forest Resource Information.* New York: United Nations.

ECE/FAO (1993b), *The UN–ECE/FAO 1990 Forest Resources Assessment. The Forest Resources of the Temperate Zones. Vol. II: Benefits and Functions of the Forest.* New York: United Nations.

ECE/FAO (1993c), *Forest Resources Assessment 1990. Tropical Countries*, FAO Forestry Paper 112. Rome.

Economist, The (1994), *Pocket World in Figures*. London: Economist Books.

FAO (1981), *Agriculture: Toward 2000*. Rome.

FAO (1985), *Yearbook of Forest Products*. Rome.

FAO (1987), *Guidelines for Forest Policy Formulation*, FAO Forestry Paper 81. Rome.

FAO (1991), *Wood and Wood Products*. Rome.

FAO (1993), *The Challenge of Sustainable Forest Management*. Rome.

FAO (1995), *Forest Resources Assessment 1990 – Global Synthesis*. Rome.

Forest Department, Sarawak (1990), *Annual Report*. Kuching.

Glesinger, E. (1949), *The coming age of wood*. New York.

Gore, R. (1989), "Extinctions", *National Geographic*, vol. 175, no. 6.

Gregory, G. Robinson (1972), *Forest Resource Economics*. New York: Ronald Press.

Grøn, A. Howard (1931), *Den almindelige Skovøkonomis Teori*. Copenhagen.

Grøn, A. Howard (1945), *Skovbrugets Driftsøkonomi III*. Copenhagen.

ICRAF (1983), *Agroforestry Systems for Small-scale Farmers*. Nairobi, Kenya.

ITTO (1990), "Status and Potential of Non-Timber Products in the Sustainable Development of Tropical Forests". Proceedings of the International Seminar Held in Kamakura, Japan, 17 November (ed. Per Wegge).

ITTO (1991), *Incentives in Producer and Consumer Countries to Promote Sustainable Development of Tropical Forests*. Pre-Project Report 22/91 (M, F, I). Oxford/Yokohama.

ITTO (1992), *Manual for Project Formulation*. Yokohama.

ITTO Mission (1990), *The Promotion of Sustainable Forest Management. A Case Study in Sarawak, Malaysia*. Yokohama.

IUCN, UNEP, WWF (1980), *World Conservation Strategy*, IUCN, Gland, Switzerland.

Jonsson, T., and Lindgren, P. (1990), *Logging Technology for Tropical Forests – for or against? Report from an ITTO Preproject*. Sweden: The Forest Operation Institute "Skogsarbeten".

Klose, Franz (1985), *A Brief History of the German Forest – Achievements and Mistakes down the Ages*. Eschborn: GTZ.

Landbruksdepartementet (1981), *Virkesproduksjonen i skogbruket*. St. meld. nr. 15 (1981–1982), Oslo.

Meadows, D. H., Meadows, D. L., Randers, J., and Behrens, W. W. (1972), *Limits to Growth*. The Club of Rome. Post Mills, VT.

Meadows, D. H., Meadows, D. L, and Randers, J. (1991), *Beyond the Limits*. Post Mills, VT.

Nersten, S., Delbeck, K., Gjølberg, R., and Hobbelstad, K. (1981), *Konsekvensanalyser for ulike investerings- og avvirkningsprogram*. Melding nr. 29, Institutt for Skogtaksasjon, Norges Landbrukshøgskole. Ås: NLH.

Nilsson, Nils-Erik (1991), "Forestry for sustainable development", Statement in ITTO Council, May. Unpublished.

Nilsson, Nils-Erik (1992), *Development of National Forest Policies and Strategies.* Jønkøping, Sweden: National Board of Forestry.

Norman, V. D. (1990), *Verden i tall.* Oslo: Economist Books, Dagens Næringsliv Forlag A/S.

Palmberg-Lerche, C. (1993), "Present status of forest plantations and tree improvement in the Americas, with special reference to tropical America". Proc. 1st Pan-American Forestry Congress/7th Brazilian Forestry Congress.

Peck, T. J. (1991), "Medium-term trends and prospects for the consumption of forest products". Proceedings of the 10th World Forestry Congress, Paris.

Poore, D. (1989), *No Timber without Trees.* London: Earthscan Publications.

Prats Llaurado, J., and Speidel, I. (1981), *Public Forestry Administrations in Latin America.* FAO Forestry Paper 25. Rome.

Primack, R. B., and Hall, P. (1992), "Biodiversity and forest change in Malaysian Borneo". Workshop paper. International Workshop on Ecology, Conservation and Management of Southeast Asian Rainforests, Kuching, Sarawak, Malaysia, October.

Seip, H. K. (1993), *Sustainable Multiple Use of Hill Forest in Sarawak, Malaysia.* Report to the Government of Sarawak on an ITTO pre-project.

Sharma, N. P., et al. (1992), *Managing the World's Forests.* Dubuque, Iowa: Kendall/Hunt Publishing Company.

Solberg, B., and Skaar, R. (1986), *A Technical and Socio-Economic Study of Skidding with Oxen in Malawi.* Ås: NORAGRIC, The Agricultural University of Norway.

Sundberg, U. (1981), *Level of Mechanization in Forest Operations,* FO: MISC/81/6. Rome: FAO.

Tomter, S. M. (1993), *Skog 1993. Statistics of Forest Conditions and Resources in Norway.* Ås: Norwegian Institute of Land Inventory.

UNCED (1992a), *Combatting Deforestation.* Prep. Com. 2/3-3/4–92. Conference paper, Rio de Janeiro.

UNCED (1992b), *Agreements on Environment and Development.* Conference paper, Rio de Janeiro.

WCED (1987a), *Our Common Future.* Oxford: Oxford University Press.

WCED (1987b), *Energy 2000.* London: Zed Books.

Westoby, Jack (1987), *The Purpose of Forests.* Oxford: Basil Blackwell.

Westoby, Jack (1989), *Introduction to World Forestry.* Oxford: Basil Blackwell.

Wilkinson, B., Matussek, H., Salvesen, W., and Sutton, P. (1992), *International Fact & Price Book 1992.* Brussels: PPI Pulp & Paper International.

Worldwatch Institute (1991), *State of the World.* Washington, D.C.: Worldwatch Institute.

Index